Introduction to Semiconductor Lasers for Optical Communications

Introduction to Semiconductor Lasers for Optical Communications

Editor

Anik Kumar

scitus
academics

Introduction to Semiconductor Lasers for Optical Communications

Edited by **Anik Kumar**

Printed in 2017

ISBN: 978-1-68117-221-7

Library of Congress Control Number: 2015936581

Contents

vi

Preface

Even in optical communications, lasers are used in different ways, from metropolitan links using directly modulated devices to 100 Gb/s transmission systems incorporating advanced detection and modulation schemes. In this book, I introduce semiconductor lasers from an operational perspective to those who have a background in engineering or optics, but no familiarity with lasers. The objective here is to present semiconductor lasers in a way that is both accessible and interesting to advanced undergraduate students and to first-year graduate students. The target audience for this book is someone who is potentially interested in careers in semiconductor lasers, and the decision of what topic to cover is driven both by the importance of the topic and how fundamental it is to the whole field.

Editor

Modeling of Millimeter-wave Modulation Characteristics of Semiconductor Lasers under Strong Optical Feedback

Ahmed Bakry

Department of Physics, Faculty of Science, King Abdulaziz University, Jeddah 21589, Saudi Arabia

ABSTRACT

This paper presents modeling and simulation on the characteristics of semiconductor laser modulated within a strong optical feedback (OFB-) induced photon-photon resonance over a passband of millimeter (mm) frequencies. Continuous wave (CW) operation of the laser under strong OFB is required to achieve the photon-photon resonance in the mm-wave band. The simulated time-domain characteristics of modulation include the waveforms of the intensity and frequency chirp as well as the associated distortions of the modulated mm-wave signal. The frequency domain characteristics include the intensity modulation (IM) and frequency modulation (FM) responses in addition to the associated relative intensity noise (RIN). The signal characteristics under modulations with both single and two mm-frequencies are considered. The harmonic distortion and the third order intermodulation distortion (IMD3) are examined and the spurious free dynamic range (SFDR) is calculated.

INTRODUCTION

The radio over fiber (RoF) technology has recently attracted considerable attention as an integration of wireless and optical systems and consequently as a solution to enhance the communication bandwidth to support integrated services [1]. This technology inherently combines the advantage of enormous bandwidth of optical fiber and the flexibility of wireless access technologies to deliver wireless RF signals directly from the central station to simplified base stations. The mm-wave bands are utilized to meet the demand for higher signal bandwidth and to overcome the frequency jamming in the RoF-based wireless networks [2]. However, enhancing the bandwidth of RoF links with directly modulated semiconductor lasers to the mm-wave band is limited by the available modulation bandwidth of semiconductor lasers. Increasing the differential gain is an efficient technique to increase the modulation bandwidth [3]. Modulation up to a frequency of 25 GHz was achieved [4]; however a record of 40 GHz response is challenging. In addition, at frequencies near the relaxation oscillation frequency, the laser noise and distortion increase [5]; therefore the usable bandwidth for directly modulated analog links is less than the possible 3 dB bandwidth.

The technique of injection locking has been reported by several groups to enhance the modulation bandwidth of laser diodes [6–8]; high-frequency modulation beyond 40 GHz was demonstrated [9, 10]. External OFB has been shown to be an alternative and cost-effective technique to increase the modulation bandwidth of semiconductor lasers, depending on appropriate choices of the system parameters [11–16]. Narrow-band high-frequency modulation over 40 GHz has been achieved in quantum well lasers under OFB [17]. Semiconductor lasers with OFB display rich chaotic dynamic behaviors, including period-1 oscillations, period doubling, quasiperiod, and routes to chaos due to variations in the phase of the reinjected field into the laser cavity [15, 18]. Under strong OFB, the frequency of the induced oscillations could be comparable to a resonance frequency of the external cavity [19]. Most recently the author's group [20] has newly reported using strong OFB to boost the modulation frequencies over an ultrahigh frequency passband over 55 GHz and has shown improvement of the gain of a corresponding RoF link by about 20 dB. Such enhancement in the IM response over an ultrahigh frequency passband was attributed as a type of photon-photon resonance due to coupling of oscillating modes in the coupled cavity [13, 21, 22]. It is obtained when the nonmodulated laser keeps operation in CW under strong OFB. The authors pointed out also that the noise factor of such mm-wave RoF links improves nearly by 20 dB in the regime of small-signal modulation and 10 dB under large-signal modulation [23].

In this paper, we introduce comprehensive investigation on the mm-wave modulation characteristics and noise of semiconductor lasers with a short-external cavity and strong OFB. We newly discuss the FM response and the frequency chirp associated with this ultrahigh speed IM. Because the signal distortion and the dynamic range of the laser are critical issues in the modulation of semiconductor lasers and the optical analog links [24,25], we also examine the harmonic distortions in the mm-wave modulated laser signal. We consider modulation of the laser with single mm-frequency and study the associated second order harmonic distortion (2HD). We study also the modulation performance of the laser modulation using two adjacent mm-frequencies and measure the associated IMD3 and the corresponding SFDR. The present study is based on applying a strong OFB rate equation model, in which OFB is treated as time delay of OFB with roundtrips (multiple reflections) in an external cavity [4]. The noise content of the mm-wave modulated signal is characterized by the frequency spectrum of RIN. The SFDR, which is defined as the dynamic range at the modulation power when the system noise floor is equal to the distortion noise, is measured by the power of the fundamental frequency component, noise floor, and the power in the IMD3 component [26]. We compare the obtained findings with those of a solitary laser when modulated at the carrier-photon resonance (relaxation) frequency, which is the most practical frequency regime of the solitary laser at which the IM is most enhanced. We apply the model to a high-speed DFB laser with a modulation bandwidth of about 25 GHz [4]. We show that, compared with the solitary laser modulated at the relaxation frequency, the present modulated signal was shown to have 5 dB lower distortions, 10 dB/Hz lower RIN, and 2 dB/Hz$^{3/2}$ higher SFDR.

In the next section, we introduce the time-delay rate equation model of analyzing OFB with sinusoidal modulation and the associated noise. The numerical procedures of the present model are given in Section 3. In Section 4, we present the simulated IM and FM responses and noise characteristics under single- and two-tone modulations. Finally, the present work is concluded in Section 5.

THEORETICAL MODEL

The dynamics and noise of semiconductor under both IM and external OFB are described by the following time-delay rate equations of the carrier number N(t), photon number S(t), and optical phase θ(t) [23]:

$$\frac{dN}{dt} = \frac{1}{e}I(t) - \frac{N}{\tau_s} - \frac{av_g}{V}\frac{N - N_g}{1 + \varepsilon S}S + F_N(t),$$
(1)

$$\frac{dS}{dt} = \left[\Gamma\frac{av_g}{V}\frac{N - N_g}{1 + \varepsilon S} - G_{th}\right]S + C\frac{N}{\tau_s} + F_S(t),$$
(2)

$$\frac{d\theta}{dt} = 2\pi\Delta v(t) = \frac{1}{2}\left(\alpha\Gamma\frac{av_g}{V}(N - N_{th}) - \frac{v_g}{L_D}\varphi\right) + F_\theta(t),$$
(3)

Where Dn(t) is the frequency chirp induced by the instantaneous variation of the optical phase due to variation in S(t) and N(t). In (2), G_{th} is the threshold gain under OFB and is determined by the photon lifetime t$_p$ in the laser cavity of length L$_D$ and refractive index,n$_D$

$$G_{th} = \frac{1}{\tau_p} - \frac{v_g}{L_D}\ln|U(t - \tau)|.$$
(4)

In the above equations, a is the differential gain coefficient, V$_g$ is the group velocity in the active layer of length L_D, Γ is the confinement factor, α is the linewidth enhancement factor, τ_s is the spontaneous emission lifetime, ε is coefficient of gain suppression, N_g is the electron number at transparency, and N_{th} is the electron number at threshold. In (4), $U(t - \tau)$ is an OFB function that describes the time delay of laser radiation due to roundtrips (i.e., multiple reflections) in the external cavity (of length L_{ex} and refractive index n_{ex}) formed between the laser front facet (of reflectivity R_f) and the external mirror (R_{ex}) [27, 28],

$$U(t - \tau) = |U(t - \tau)| e^{-j\varphi}$$

$$= 1 - \sum_{p=1} (K_{ex})^p \left(\frac{R_f}{1 - R_f} \right)^{p-1} e^{-j\omega\tau} \frac{S(t - p\tau)}{S(t)} \frac{e^{j\theta(t-p\tau)}}{e^{j\theta(t)}},$$

$$(5)$$

$$\varphi = -\tan^{-1} \frac{\text{Im}\{U(t - \tau)\}}{\text{Re}\{U(t - \tau)\}} + n\pi \quad n : \text{integer}$$

$$(6)$$

with ω being the angular frequency of the laser emission and $\tau = 2n_{ex}L_{ex}/c$ as the roundtrip time. The strength of OFB is measured by the coupling coefficient K_{ex}, which is determined by the ratio between R_{ex} and R_f [27, 28]:

$$K_{ex} = (1 - R_f) \sqrt{\eta \frac{R_{ex}}{R_f}},$$

$$(7)$$

where η is the external coupling efficiency of the injected light into the laser cavity. In (6) n is an integer and is chosen to vary continuously for time evolution, because the solution of arc tangent is limited in the range of $-\pi/2$ to $\pi/2$ in the computer work. At a given time t, the phase difference between the time-delayed (externally injected) field and the field inside the laser cavity is given by $\theta(t - m\tau) - (t)$, which is equal to zero or π in the cases of in-phase and out-of-phase conditions. The injection current (t) is composed of a bias component I_b and a sinusoidal component of amplitude I_m and frequency f_m:

$$I(t) = I_b + I_m \sin(2\pi f_m t).$$

$$(8)$$

The modulation depth is given as $m = I_m/I_b$. The last terms $F_N(t)$, $F_S(t)$, and $F_\theta(t)$ in rate equations (1)–(3) are Langevin noise sources with zero mean values and are added to the equations to account for intrinsic fluctuations of the laser [28]. These noise sources are assumed to have Gaussian probability distributions and to be δ-correlated processes [28]. The frequency content of intensity fluctuations is measured in terms of RIN, which is calculated from the fluctuations $\delta S(t) = S(t) - \overline{S}$ in $S(t)$, where $\overline{S}$ is the time-average value of $S(t)$. Over a finite time T, RIN is given as [29]

$$\text{RIN} = \frac{1}{\overline{S}^2} \left\{ \frac{1}{T} \left| \int_0^T \delta S(t) \, e^{-j2\pi f\tau} d\tau \right|^2 \right\},$$

(9)

where f is the Fourier noise frequency.

NUMERICAL CALCULATIONS

Rate equations (1)–(3) are solved numerically by the 4th order Runge-Kutta method using a time integration step as short as 0.2 ps to allow simulation of the very high speed modulated signal. Five roundtrips, $p=1 \rightarrow 5$, are counted in the calculations. At each integration instant, the noise sources (t), $F_s(t)$, and $F_\theta(t)$ are generated following the technique developed in [30] using a set of three uniformly distributed random numbers generated by the computer. In the simulations, we use the numerical values listed in Table 1, which correspond to single-mode quantum-well DFB laser [27]. This laser has a threshold current of $I_{th} = 10$ mA. The laser is assumed to be biased above threshold, $I_b = 5I_{th}$. We adjust the length of the external cavity to be $n_{ex}L_{ex} = 0.25$ cm, which corresponds to an external-cavity resonance frequency spacing ~60 GHz. The fast Fourier transform (FFT) is used to simulate the frequency content of the modulated laser signal. The IM response and the associated FM response are calculated numerically, respectively, as

$$\text{IM} - \text{repsonse} = \frac{a_1(f_m)}{a_1(f_m \rightarrow 0)},$$

(10)

$$\text{FM} - \text{response} = \frac{b_1\left(f_m\right)}{I_m},$$

(11)

where $a_1(f_m)$ and $b_1(f_m)$ are the fundamental harmonics of the FFT spectra of the laser intensity and frequency at the modulation frequency f_m.

Table 1: Definition and numerical values of the solitary high-speed laser parameters

Symbol	Definition	Value
λ	Wavelength	1.55 µm
V	Active layer volume	$3 \times 10^{-17}\,\text{m}^3$
Vg	Group velocity	$8.33 \times 10^7\,\text{m/s}$
LD	Active layer length	120 µm
a	Differential gain coefficient	$8.25 \times 10^{-12}\,\text{m}^2$
Ng	Carrier number at transparency	$3.69 \times 10^7\,\text{m}^{-3}$
	Linewidth enhancement factor	3.5
Γ	Confinement factor	0.15
τ_p	Photon lifetime	1.69 ps
τ_s	Spontaneous emission lifetime	776 ps
R_f	Front facet reflectivity	0.2
R_b	Back facet reflectivity	0.6
sp	Spontaneous emission factor	3×10^{-5}
ε	Nonlinear gain suppression factor	$2.77 \times 10^{-23}\,\text{m}^3$

RESULTS AND DISCUSSION

Modulation Response of the Solitary Laser

Analog modulation of the injection current leads to variation in the injected carrier density into the active region, which results in a variation of the laser frequency. Figure 1 plots both the simulated IM and FM responses of the solitary laser for when the modulation is as weak as the modulation depth which is $m = 0.1$. The FM response has a maximum at the relaxation frequency $f_r \sim 15$ GHz, which is slightly above the IM response at $f_p = 14$ GHz [31].The IM response has a 3dB modulation bandwidth of $f_{3dB} = 25$ GHz. The modulation response spectrum can be understood as follows [32]. When f_m is much lower than f_r, the injected carriers follow the change in the injection current resulting in a flat response. Around the resonance frequency, the charge carriers interact with the photons with phase synchronization, which results in the laser resonance and the peaked response. The declining part of the modulation response is because the phase of the photon field lags behind that of the injection current. When f_m increases beyond f_r, the electron and photon fields tend to become more and more out of phase, resulting in damping of the relaxation oscillations and reduction in the IM response.

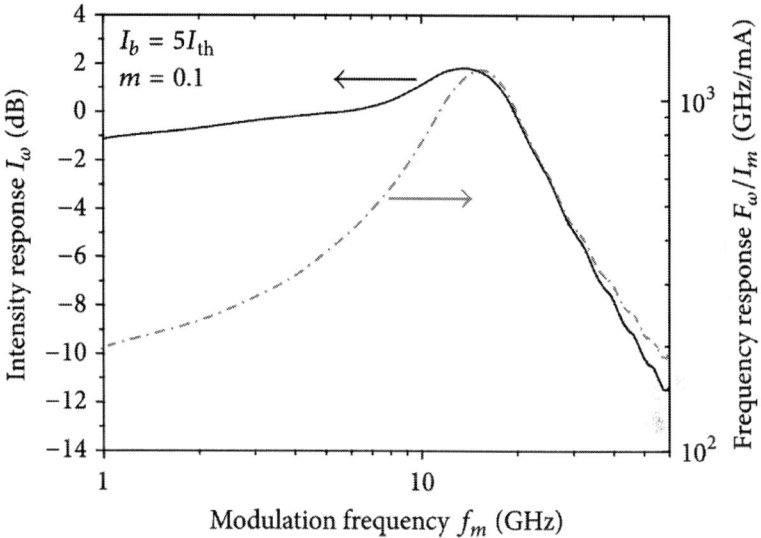

Figure 1: The intensity and frequency modulation responses of the solitary laser when $I_b = 5I_{th}$ and $m = 0.1$.

Modulation Response under OFB

The present case of a semiconductor laser with a short cavity is characterized by a frequency ratio $f_{ex}/f_r > 1$, which corresponds to a period-doubling route-to-chaos [18, 33, 34]. In the regime of strong OFB, the frequency of the possible oscillations may reach the external-cavity resonance frequency $f_{ex} = n_{ex}L_{ex}/c$. In Figure 2(a), we plot examples of the numerical IM responses of the laser under strong OFB that are characterized by resonance enhancement over a mm-waveband. In Figure 2(b), we plot the corresponding FM responses. The shown IM responses are simulated for two short-external cavities with lengths of $L_{ex} = 0.25$ and 0.30 mm, which correspond to external-cavity resonance frequencies of 60 and 30 GHz, respectively. In this case the modulation depth is $m = 0.1$ and the OFB is as strong as $K_{ex} = 1.45$. Our simulation showed that the nonmodulated laser diode operates in CW in this level of OFB, where the injected delay light is nearly in phase with the optical field in the laser cavity. The figure shows that the IM response drops under the –3 dB level at the frequencies of $f_m = 8$ and 6 GHz when $L_{ex} = 0.25$ and 0.30 mm, respectively, which are much lower than $f_{3\,dB}$ of the solitary laser. In the high-frequency regime the IM response is enhanced over the mm-wave bassbands of (54.4 and 56.5 GHz) and (45 and 46.6 GHz) centered at the frequencies f_m = 55.8 and 46 GHz when L_{ex} = 0.25 and 0.30 mm, respectively. These frequency bands are much higher than $f_{3\,dB}$ of the solitary laser. The IM enhancement over the IM response of the solitary laser is as large as 5.4 dB and 6.5 dB, respectively, which may be due to higher degree of phase-matching between the coupling OFB and the optical field in the laser cavity. Similar behavior of the narrow-band enhancement of the IM was reported by Troppenz et al. [16] around 40 GHz. Figure 2(b) shows that, contrary to the case of the solitary laser, the peaks of the FM responses occur at the same peak frequencies of the IM responses. The FM responses are lower than that of the solitary laser at frequencies lower than the mm-frequency passbands but are enhanced within these frequency bands. The amplitudes of these FM responses are ~3.2 and 5.0 times larger than that of the solitary laser when L_{ex} = 0.25 and 0.30 mm, respectively.

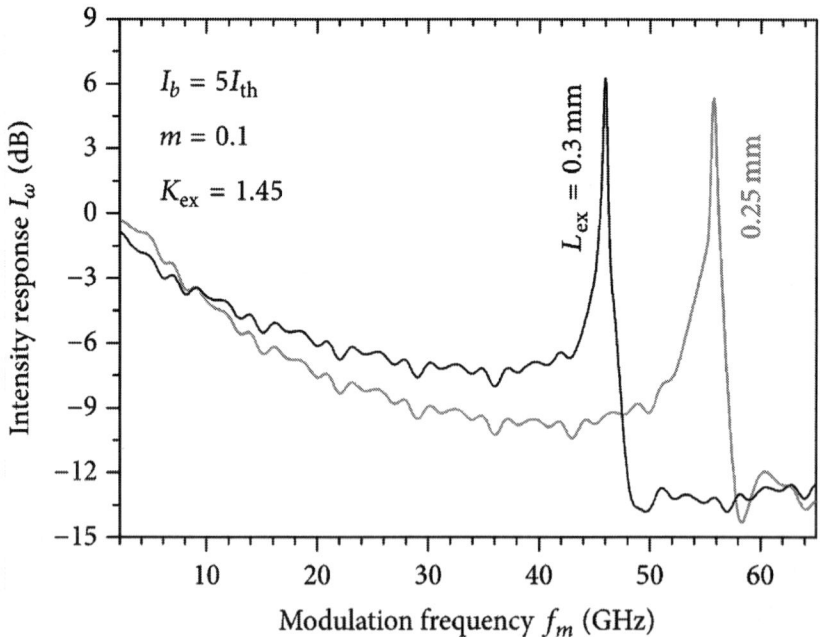

(a)

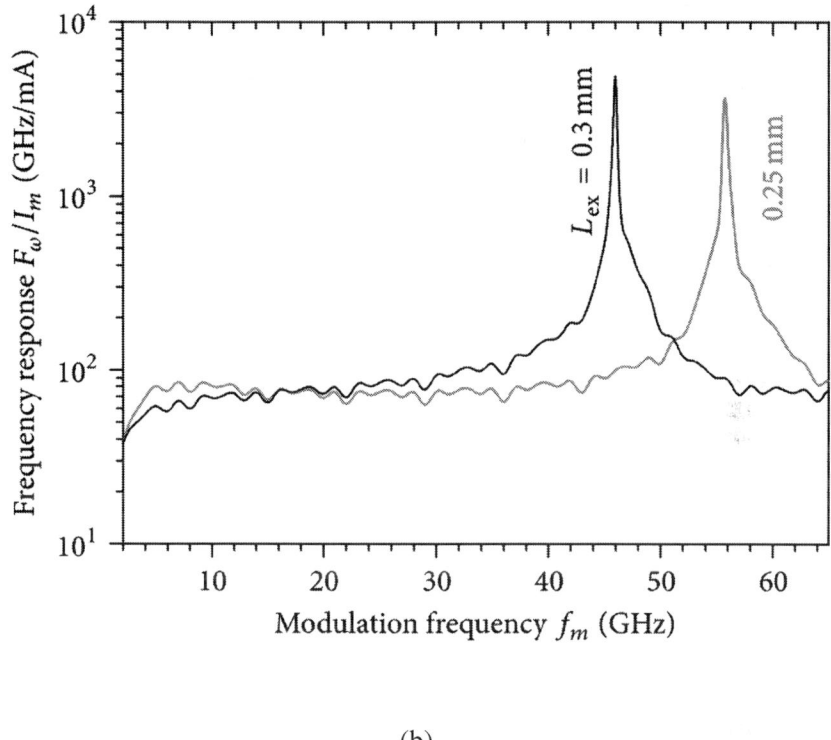

(b)

Figure 2: The intensity modulation responses of the laser under OFB with K_{ex} = 1.45 when L_{ex} = 0.25 and 0.30 mm with m = 0.1

This mm-narrow band enhancement of the modulation response can be attributed to coupling between the resonance modes of the external cavities because of the carrier pulsation in the laser cavity at the beating frequency f_{ex}. This carrier pulsation is induced by the modulating current signal as indicated by (8) and the rate (1) of the injected carrier number . This resonance is induced by optical modes and is different from the conventional carrier-photon resonance, which occurs around the relaxation frequency of the laser. Therefore, this resonance is referred to as "photon-photon resonance" [13]. Similar effect is observed in vertical-cavity surface-emitting lasers (VCSELs) coupled to a transverse cavity [22], in which the photon-photon resonance is induced by transverse oscillating modes. Because the shown IM responses have an unused frequency between the frequency f_{3dB} and the enhanced mm-narrow bands, this modulation enhancement is not favored for applications in telecommunications. It is interesting for applications such as the mm-wave RoF networks that require only a narrow bandwidth centered at a millimeterwave. To recover the modulation response over this unused gap and achieve flat IM responses, dispersive techniques may be needed at the end mirrors to obtain and allocate the photon-photon resonance at mm-frequency and achieve a wide carrier-photon resonance [21].

Modulation Performance in the mm-Frequency Band

As shown above, when the laser is subjected to strong OFB, the modulation frequency is enhanced over a narrow mmfrequency band that can be very close to the external cavity resonance frequency f_{ex}. In this section, we characterize the laser modulation at the peak-frequencies of f_m = 55.8 and 46 GHz, which correspond to the external-cavity lengths of 0.25 and 0.30 mm, respectively, and compare the results with the modulation characteristics of the solitary laser when modulated at the carrier-photon resonance frequency f_r. These characteristics include the waveforms of the signal power (t) and frequency chirp $\Delta v(t)$ and the associated harmonic distortion and RIN. We also characterize the modulation response to two-tone sinusoidal modulation and calculations of the corresponding intermodulation distortion and SFDR.

Single-Tone Modulation

Figures 3(a) and 3(b) plot the time variations of the photon number $S(t)$ and the associated frequency chirp $\Delta v(t)$, respectively, of the mm-frequency modulated laser signals that correspond to Figure 2. For comparison, the signal characteristics of the solitary laser modulated at $f_m=f_r$ are also plotted in the figures. Figure 3(a) indicates that the modulated signals are of the period-1 type oscillation. The oscillation amplitude of the modulated signal when L_{ex} = 0.30 mm is larger than that when L_{ex} = 0.25 mm, which agrees with the higher IM enhancement shown in Figure 2(a). The oscillation frequency of the signals is 15, 55.8, and 46 GHz of the solitary laser and laser with external cavities of lengths L_{ex} = 0.25 and 0.30 mm, respectively. The modulated signals under OFB are almost sinusoidal, whereas the modulated signal of the solitary laser with $f_m = f_r$ tends to be clipped. The fast Fourier transform (FFT) analysis of these signals indicates that the mm-wave modulated signals have 2nd order harmonic-order distortion of 2HD = −7.4 and −6.0 dB when L_{ex} = 0.25 and 0.30 mm, respectively, which are lower than that of the solitary laser (−2.87 dB). The 2HD is calculated as [35]

$$2HD = 10 \log_{10} \frac{a_2}{a_1},$$

(12)

where a_1 and a_2 are the FFT components at the fundamental frequency f_m and its second order harmonic, respectively.

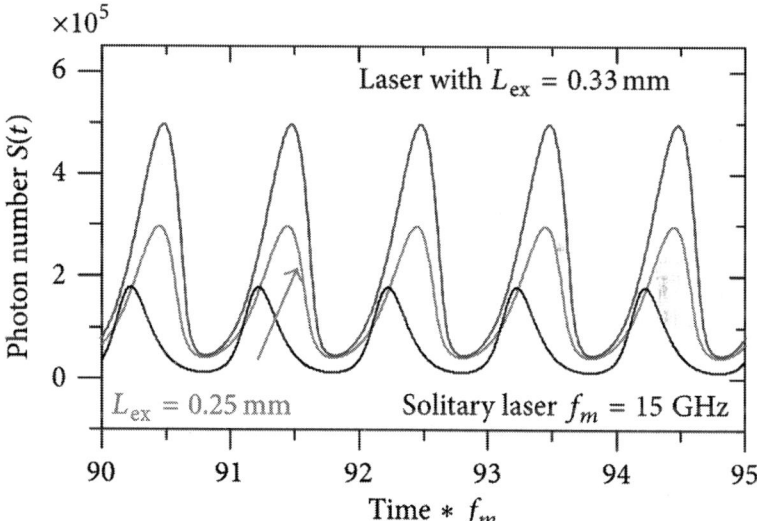

(a)

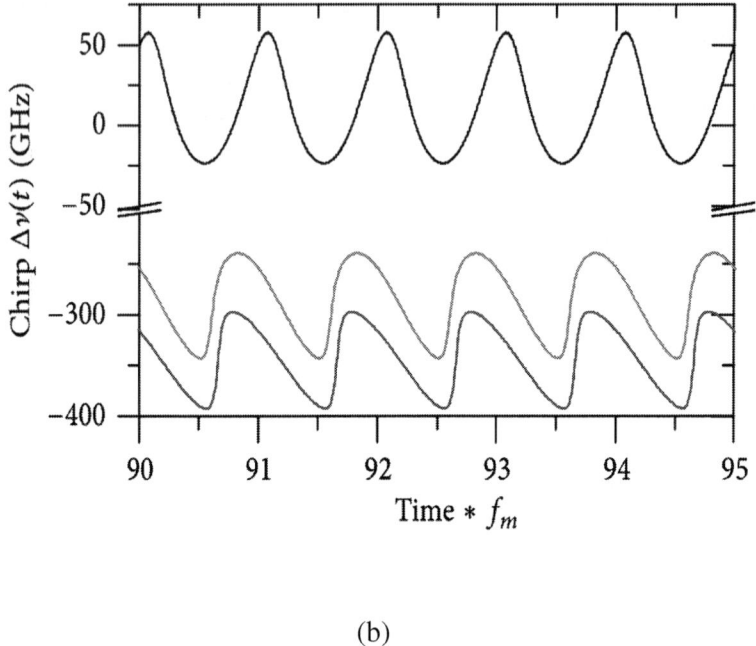

(b)

Figure 3: Modulation characteristics of the laser under OFB with L_{ex} = 0.25 mm and f_m = 55.8 GHz, with L_{ex} = 0.30 mm and f_m = 46 GHz, and the solitary laser with f_m = 15 GHz: (a) modulated waveform S(t) and (b) frequency chirp Δv(t).

On the other hand, Figure 3(b) shows that both the mmmodulated lasers under strong OFB with L_{ex} = 0.25 mm and 0.30 mm are red-shifted and the depth of the frequency chirp $\Delta v(t)|_{max}$ is 103 and 95 GHz, respectively, while the 15 GHzmodulated solitary laser is blue shifted and $\Delta v(t)|_{max}$ = 57 GHz. These results are consistent with relative amplitudes of the FM response of the solitary laser in Figure 1 and the lasers under OFB with L_{ex} = 0.25 and 0.30 mm in Figure 2. The frequency red shift of the laser with enhanced IM response is because the strong OFB results in a decrease in the carrier number (t) under the threshold level N_{th} [27], as indicated from the rate equation (3).

The comparison of the depth of the frequency chirp $\Delta v(t)|_{max}$ between the modulated signal of solitary laser and those of the laser under OFB when L_{ex} = 0.25 and 0.30 mm is examined over a wide range of the modulation index m as given in Figure 4. Figure 4 shows that the depth of the frequency chirp $\Delta v(t)|_{max}$ increases almost linearly with the increase in m for the three cases. Over the entire range of m, $\Delta v(t)|_{max}$ of the 46 and 55.8 GHz-modulated lasers under OFB are almost 30–70 GHz and 40 GHz larger than that of the 15 GHz-modulated solitary laser.

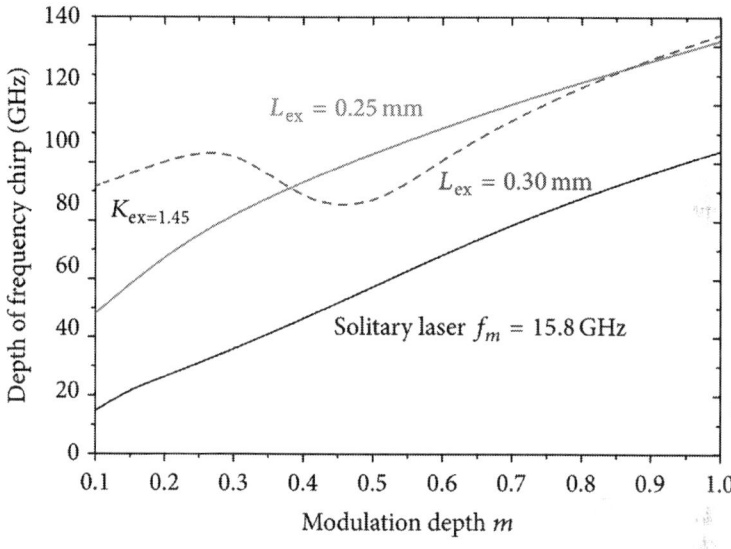

Figure 4: The frequency chirp $\Delta v(t)$ associated with intensity modulation as a function of the modulation depth for both the ultrahigh frequency modulated laser under OFB with $L_{ex} = 0.25$ and 0.30 mm and the solitary laser with $f_m = fr$.

We examine the noise content of the modulated signal in terms of the spectral characteristics of RIN. In Figure 5, we plot the frequency spectra RIN of the mm-frequency modulated laser under strong OFB. Figure 5(a) plots the RIN spectrum of the solitary laser when modulated at the relaxation frequency $f_m = f_r$, while Figures 5(b) and 5(c) plot the RIN spectra of the 46 GHz and 55.8 GHz-modulated laser under OFB from external cavities with $L_{ex} = 0.30$ mm and 0.25 mm, respectively. The figures show that the RIN spectra have sharp peaks at the corresponding modulation frequency f_m and at the higher harmonics. The low-frequency part of the RIN spectrum is almost flat (white noise). The level of this low-frequency noise, LF-RIN, is an inverse measure of the signal to noise ratio of the modulated signal [36]. The figures indicate that LF-RIN of the mm-frequency modulated laser is almost one-order of magnitude lower than that of the 15 GHz-modulated solitary laser. That is, compared with the solitary laser modulated at the relaxation frequency, the laser subjected to strong OFB and modulated within the mmfrequency band with enhanced modulation characterized by a signal with lower amplitude of intensity fluctuations.

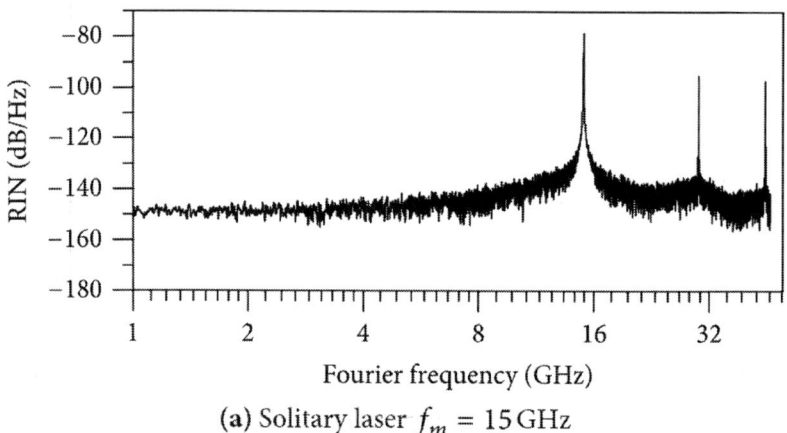

(a) Solitary laser $f_m = 15\,\text{GHz}$

(a)

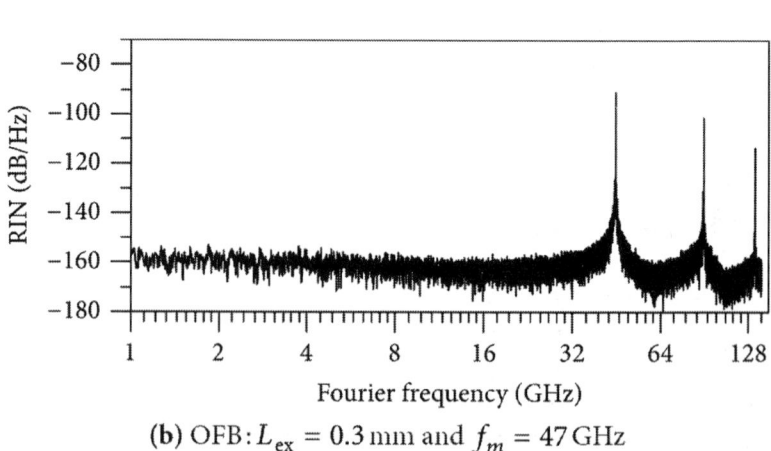

(b) OFB: $L_{ex} = 0.3\,\text{mm}$ and $f_m = 47\,\text{GHz}$

(b)

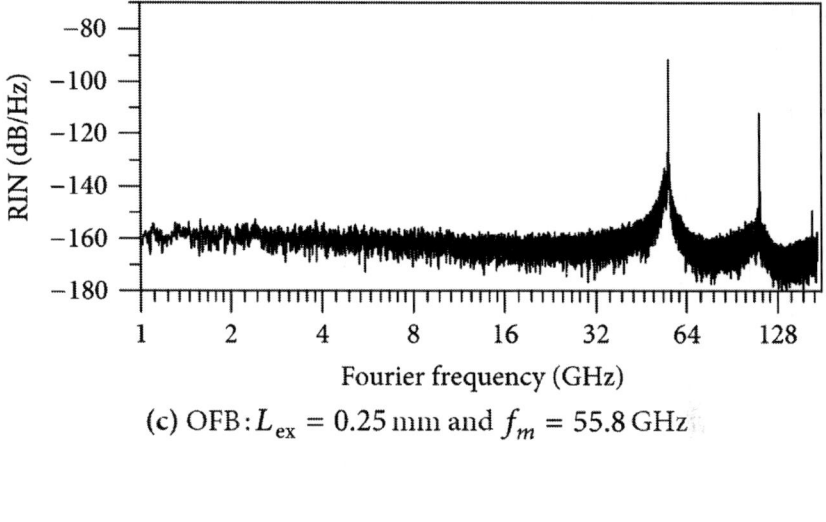

(c) OFB: $L_{ex} = 0.25$ mm and $f_m = 55.8$ GHz

(c)

Figure 5: Frequency spectra of RIN of (a) solitary laser modulated at $f_m = f_r$, (b) laser under OFB with $L_{ex} = 0.30$ mm and $f_m = 46$ GHz, and (c) laser under OFB with $L_{ex} = 0.25$ mm and $f_m = 55.8$ GHz..

Two mm-Tone Modulation Characteristics

In the case of modulation with two mm-frequencies f_{m1} and f_{m2} with $f_{m1} < f_{m2}$, the injection current $I(t)$ in (1) is given by

$$I = I_b \{1 + m [\cos(2\pi f_{m1} t) + \cos(2\pi f_{m2} t)]\}.$$

(13)

Such two-tone modulation is important for several applications, such as multichannel RF-frequency division multiplexed transmission of analog or microwave signals [37]. However, this modulation is often associated with intermodulation distortion, which occurs when the nonlinearity of the laser causes undesired outputs at sum and difference frequencies. The IM3 of two closely spaced carrier frequencies (at f_2 and $f_2 + \Delta f$) is of particular interest [5]. Figure 6(a) plots the two-tone modulation response of the laser under OFB when modulated at f_{m1} = 55.8 GHz and $f_{m2} + \Delta f$ using the frequency spacing $\Delta f = 10$ MHz. The figure corresponds to the modulation depth $m = 0.5$. The figure shows appearance of the 3rd order intermodulation components at

$f_{m1} - \Delta f$ and $f_{m2} + \Delta f$ in addition to the fundamental harmonics at f_{m1} and f_{m2}. IMD3 is defined as the ratio, in dB, of the amplitude of the third order intermodulation component to that of the fundamental component [26]:

$$\text{IMD3} = 10 \log_{10} \frac{a_{f_{m2} + \Delta f}}{a_{f_{m2}}}.$$

(14)

Figure 6(b) plots IDM3 as a function of the modulation depth m. The figure shows that IMD3 increases with the increase in m. The slope of such increase is large in the regime of smallsignal modulation and decreases with the increase in m. The figure indicates that IMD3 ranges between −14.5 and −8 dB. In the figure, we also compare the IMD3 values with those of the solitary laser when modulated at the carrier-photon resonance frequency $f_{m1} = fr$. As shown in the figure, IMD3 of the solitary laser is little lower than that of the laser under strong OFB up to $m = 0.2$. For modulation with larger signals, IMD3 of the solitary laser becomes larger and the differences reach 3 dB when $m = 1.0$. These values of IMD3 limit the performance of the laser-based RoF links. This limitation is measured by SFDR. SFDR is determined by three quantities, namely, the power of the fundamental frequency component, noise floor, and the power in the IMD3 component [26]. The noise floor is determined from the RIN spectrum of the free-running laser when $I_b = 5I_{th}$. Figure 6(c) plots the output powers of the fundamental signal and the 3rd order intermodulation power and the noise floor versus the input electrical power. A linear fit is made to the plotted data and the SFDR is extracted as SDFR = 83 dB/Hz3/2. The calculated value of SFDR of the solitary laser when modulated at the $f_m = f_r$ was found to be 2 dB/Hz$^{3/2}$ lower.

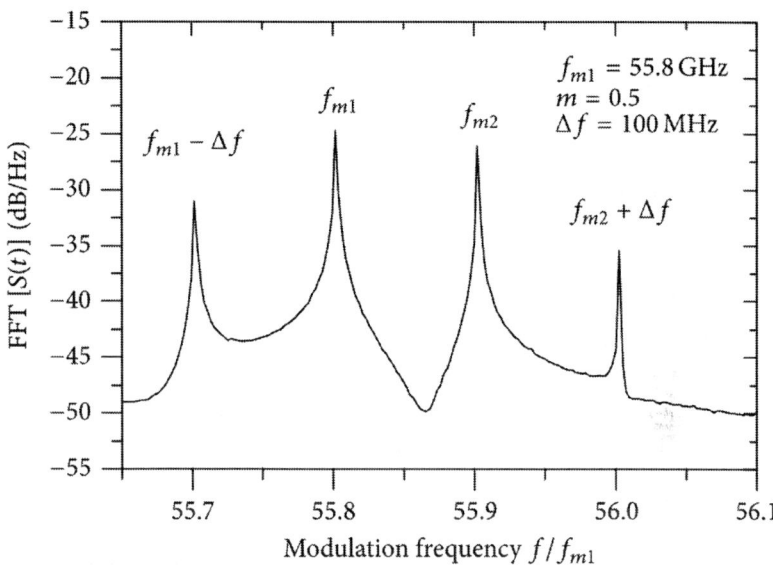

(a)

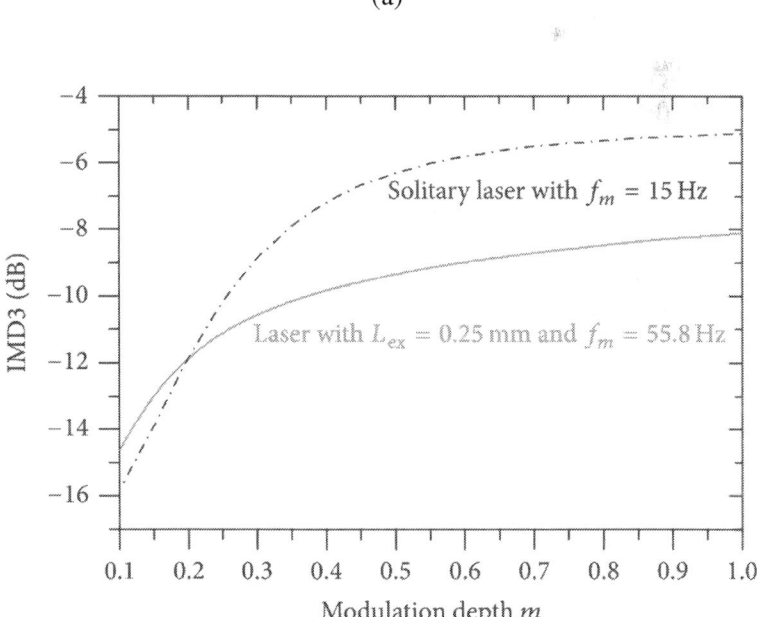

(b)

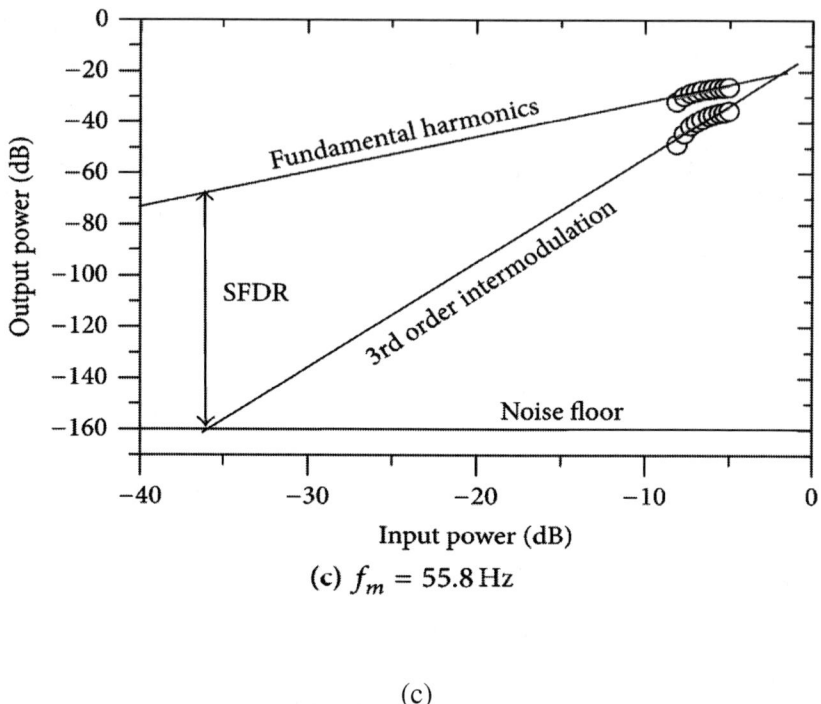

(c) $f_m = 55.8\,\text{Hz}$

(c)

Figure 6: Characteristics of two-tone modulation with $f_{m1} = 55.8$ GHz and $\Delta f = 100$ MHz: (a) FFT power spectrum showing the intermodulation components at $f_{m1} - \Delta f$ and $f_{m2} + \Delta f$, (b) influence of modulation depth m on the IMD3, and (c) SFDR determination, SFDR = 73 dB/Hz3/2.

CONCLUSIONS

We presented the modeling of mm-frequency modulation characteristics of semiconductor lasers under strong OFB. The study was based on the theoretical modeling fully handling the strong OFB regime as time delay of laser light due to round-trips in the external cavity. We analyzed the signal distortions and noise associated with both single and two-tone modulations. We show that the enhanced IM response under strong OFB is due to photon-photon resonance resulting from coupling of oscillating modes in the external cavity. When the beating frequency of these coupled modes matches the frequency of the modulating electrical signal, the modulation response reveals resonance over a narrow-frequency band. A key parameter to achieve this mm-wave photon-photon resonance is to modulate the laser when it keeps stable operation in CW under strong OFB, where the injected delay light becomes in phase with the optical field in the laser cavity. Within this mmfrequency passband with enhanced IM response, the laser emits period-1 oscillations with low harmonic distortion. The LF-RIN level increases very little with the increase in the modulation depth. Under modulation with two adjacent mm-frequencies, IMD3 increases with the increase in the modulation index m, ranging between −14.5 and −8 dB. Compared with the solitary laser modulated at the relaxation frequency, the present modulated signal was shown to have 10 dB/Hz lower RIN and 2 dB/Hz3/2 higher SFDR..

ACKNOWLEDGMENT

This work was funded by the Deanship of Scientific Research (DSR), King Abdulaziz University, Jeddah, under grant No. (130-134-D1435). The author, therefore, acknowledges with thanks DSR technical and financial support.

REFERENCES

1.   R. Llorente, S. Walker, I. T. Monroy et al., "Triple-play and 60-GHz radio-over-fiber techniques for next-generation optical access networks," in Proceedings of the 16th European Conference on Networks and Optical Communications, pp. 16–19, July 2011.

2.   T. Kuri, K.-I. Kitayama, A. Stöhr, and Y. Ogawa, "Fiber-optic millimeter-wave downlink system using 60 GHz-band external modulation," Journal of Lightwave Technology, vol. 17, no. 5, pp. 799–806, 1999. ··

3.   S. Weisser, E. C. Larkins, K. Czotscher et al., "Damping-limited modulation bandwidths up to 40 GHz in undoped short-cavity $In_{0.35}Ga_{0.65}As$-GaAs multiple-quantum-well lasers," IEEE Photonics Technology Letters, vol. 8, no. 5, pp. 608–610, 1996. ·

4.   K. Sato, S. Kuwahara, and Y. Miyamoto, "Chirp characteristics of 40-Gb/s directly modulated distributed-feedback laser diodes,"

Journal of Lightwave Technology, vol. 23, no. 11, pp. 3790–3797, 2005. · ·

5. W. I. Way, "Large signal nonlinear distortion prediction for a single mode laser diode under microwave intensity modulation," Journal of Lightwave Technology, vol. 5, no. 3, pp. 305–315, 1987.

6. J. Wang, M. K. Haldar, L. Li, and F. V. C. Mendis, "Enhancement of modulation bandwidth of laser diodes by injection locking," IEEE Photonics Technology Letters, vol. 8, no. 1, pp. 34–36, 1996. · ·

7. T. B. Simpson and J. M. Liu, "Enhanced modulation bandwidth in injection-locked semiconductor lasers," IEEE Photonics Technology Letters, vol. 9, no. 10, pp. 1322–1324, 1997. · ·

8. S. K. Hwang, J. M. Liu, and J. K. White, "35-GHz intrinsic bandwidth for direct modulation in 1.3-μm semiconductor lasers subject to strong injection locking," IEEE Photonics Technology Letters, vol. 16, no. 4, pp. 972–974, 2004. · ·

9. S. C. Chan, S. K. Hwangb, and J. M. Liua, "Radio-over-fiber transmission from an optically injected semiconductor laser in period-one state," Proceedings of SPIE, vol. 6468, Article ID 646811, 2007.

10. E. K. Lau, X. Zhao, H. K. Sung, D. Parekh, C. Chang-Hasnain, and M. C. Wu, "Strong optical injection-locked semiconductor lasers demonstrating > 100-GHz resonance frequencies and 80-GHz intrinsic bandwidths," Optics Express, vol. 16, no. 9, pp. 6609–6618, 2008. · ·

11. U. Feiste, "Optimization of modulation bandwidth in DBR lasers with detuned bragg reflectors," IEEE Journal of Quantum Electronics, vol. 34, no. 12, pp. 2371–2379, 1998. · ·

12. G. Morthier, R. Schatz, and O. Kjebon, "Extended modulation bandwidth of DBR and external cavity lasers by utilizing a cavity resonance for equalization," IEEE Journal of Quantum Electronics, vol. 36, no. 12, pp. 1468–1475, 2000. · ·

13. M. Radziunas, A. Glitzky, U. Bandelow et al., "Improving the modulation bandwidth in semiconductor lasers by passive feedback," IEEE Journal on Selected Topics in Quantum Electronics, vol. 13, no. 1, pp. 136–142, 2007. · ·

14. Y. Senlin, "Modulation response characteristics of an optical delayed feedback semiconductor laser," inProceedings of the Symposium on Photonics and Optoelectronics (SOPO ‹11), pp. 1–4, Wuhan, China, 2011. ·

15. J. S. Lawrence and D. M. Kane, "Nonlinear dynamics of a laser diode with optical feedback systems subject to modulation," IEEE Journal of Quantum Electronics, vol. 38, no. 2, pp. 185–192, 2002. · ·

16. U. Troppenz, J. Kreissl, W. Rehbein, C. Bornholdt, B. Sartorius, and M. Schell, "40 Gbit/s directly modulated passive feedback laser," in Proceedings of the International Conference on Indium Phosphide and Related Materials (IPRM ‹08), pp. 1–4, Versailles, France, May 2008. · ·

17. R. Nagarajan, S. Levy, and J. E. Bowers, "Millimeter wave narrowband optical fiber links using external cavity semiconductor lasers," Journal of Lightwave Technology, vol. 12, no. 1, pp. 127–136, 1994. · ·

18. M. Ahmed, M. Yamada, and S. Abdulrhmann, "Numerical modeling of the route-to-chaos of semiconductor lasers under optical feedback and its dependence on the external-cavity length,"International Journal of Numerical Modelling: Electronic Networks, Devices and Fields, vol. 22, no. 6, pp. 434–445, 2009. · · ·

19. M. Ahmed and M. Yamada, "Field fluctuations and spectral line shape in semiconductor lasers subjected to optical feedback," Journal of Applied Physics, vol. 95, no. 12, pp. 7573–7583, 2004. · ·

20. M. Ahmed, A. Bakry, R. Altuwirqi, M. S. Alghamdi, and F. Koyama, "Enhancing modulation bandwidth of semiconductor lasers beyond 50 GHz by strong optical feedback for use in millimeter-wave radio over fiber links," Japanese Journal of Applied Physics, vol. 52, no. 12R, Article ID 124103, 2013. · ·

21. I. Montrosset and P. Bardella, "Laser dynamics providing enhanced modulation bandwidth," inProceedings of the 6th SPIE, Semiconductor Lasers and Laser Dynamics, vol. 91340H, p. 15, 2014.

22. H. Dalir, M. Ahmed, A. Bakry, and F. Koyama, "Compact electro-absorption modulator integrated with vertical-cavity surface-

emitting laser for highly efficient millimeter-wave modulation," Applied Physics Letters, vol. 105, Article ID 081113, 2014. ·

23. M. Ahmed, A. Bakry, R. Altuwirqi, M. S. Alghamdi, and F. Koyama, "Intensity noise in ultra-high frequency modulated semiconductor laser with strong feedback and its influence on noise figure of rof links," Journal of the European Optical Society, vol. 8, Article ID 13064, 2013. · ·

24. M. Ahmed, N. Z. El-Sayed, and H. Ibrahim, "Chaos and noise control by current modulation in semiconductor lasers subject to optical feedback," The European Physical Journal D, vol. 66, no. 5, article 141, 2012. · ·

25. R. V. Dalal, R. J. Ram, R. Helkey, H. Roussell, and K. D. Choquette, "Low distortion analogue signal transmission using vertical cavity lasers," Electronics Letters, vol. 34, no. 16, pp. 1590–1591, 1998. · ·

26. C. H. Cox III, Analog Optical Links, Cambridge University Press, New York, NY, USA, 2004.

27. S. G. Abdulrhmann, M. Ahmed, T. Okamoto, W. Ishimori, and M. Yamada, "An improved analysis of semiconductor laser dynamics under strong optical feedback," IEEE Journal on Selected Topics in Quantum Electronics, vol. 9, no. 5, pp. 1265–1274, 2003. · ·

28. S. Abdulrhmann, M. Ahmed, and M. Yamada, "New model of analysis of semiconductor laser dynamics under strong optical feedback," in 11th Physics and Simulation of Optoelectronic Devices, vol. 4986 ofProceedings of SPIE, pp. 490–501, 2003. ·

29. M. Ahmed, M. Yamada, and M. Saito, "Numerical modeling of intensity and phase noise in semiconductor lasers," IEEE Journal of Quantum Electronics, vol. 37, no. 12, pp. 1600–1610, 2001. · ·

30. M. Ahmed, "Numerical approach to field fluctuations and spectral lineshape in InGaAsP laser diodes,"International Journal of Numerical Modelling: Electronic Networks, Devices and Fields, vol. 17, no. 2, pp. 147–163, 2004. · · ·

31. T. L. Koch and J. E. Bowers, "Nature of wavelength chirping in directly modulated semiconductor lasers," Electronics Letters, vol. 20, no. 25-26, pp. 1038–1040, 1984. · ·

32. C. Y. Wu, Analysis of high-speed modulation of semiconductor lasers by electron heating [M.S. thesis], University of Toronto, Toronto, Canada, 1995.

33. Y. H. Kao, N. M. Wang, and H. M. Chen, "Mode description of routes to chaos in external-cavity coupled semiconductor lasers," IEEE Journal of Quantum Electronics, vol. 30, no. 8, pp. 1732–1739, 1994. ··

34. M. Ahmed, "Longitudinal mode competition in semiconductor lasers under optical feedback: regime of short-external cavity," Optics and Laser Technology, vol. 41, no. 1, pp. 53–63, 2009. ··

35. G. Keiser, Optical Fiber Communications, McGraw-Hill, New York, NY, USA, 2nd edition, 1991.

36. M. Ahmed, "Spectral lineshape and noise of semiconductor lasers under analog intensity modulation,"Journal of Physics D: Applied Physics, vol. 41, no. 17, Article ID 175104, 2008. ··

37. E. I. Ackerman and C. H. Cox III, "RF fiber-optic link performance," IEEE Microwave Magazine, vol. 2, no. 4, pp. 50–58, 2001. ··

Chapter 2

Theoretical Modeling of Intensity Noise in InGaN Semiconductor Lasers

Moustafa Ahmed

Department of Physics, Faculty of Science, King Abdulaziz University, P.O. Box 80203, Jeddah 21589, Saudi Arabia

ABSTRACT

This paper introduces modeling and simulation of the noise properties of the blue-violet InGaN laser diodes. The noise is described in terms of the spectral properties of the relative intensity noise (RIN). We examine the validity of the present noise modeling by comparing the simulated results with the experimental measurements available in literature. We also compare the obtained noise results with those of AlGaAs lasers. Also, we examine the influence of gain suppression on the quantum RIN. In addition, we examine the changes in the RIN level when describing the gain suppression by the case of inhomogeneous spectral broadening. The results show that RIN of the InGaN laser is nearly 9 dB higher than that of the AlGaAs laser.

INTRODUCTION

InGaN laser diodes have been the subject of considerable attention because of their applications in high density optical disc storage and optical data processing. In particular, blue-violet laser diodes operating at a wavelength around 400 nm are required for Blu-ray disc systems if the disk storage capacity is to be increased up to 25 Gbytes [1]. Much progress in the developments of violet-blue lasers has been made since the first operation of such a laser was reported by Nakamura et al. [2, 3]. Meanwhile, several groups have reported continuous-wave operation at room temperature using different fabrications methods [4–8].

A typical limiting factor for the optic-disc application is the noise in the laser emission, which may cause errors in the reading/recording processes. Semiconductor laser radiation intrinsically shows intensity and phase fluctuations which induce broadening of the spectral line. These fluctuations are associated with quantum transitions of charge carriers between the conduction and valence bands through the recombination processes of charge carriers and the processes of photon emission and absorption [9]. This intrinsic noise is unavoidable and is called quantum noise [10] or "optical shot noise" [11]. The laser noise may be amplified by other effects such as competition among the oscillating modes [12], external modulation [13], and external optical feedback [14]. These types of noise have been intensively investigated in both theory and experiment for near infrared GaAs and In P-based laser diodes [14–20]. Experimental studies on the noise of blue-violet InGaN lasers showed that the properties of noise in the blue-violet laser are not so different qualitatively from those in the infrared lasers [11]. However, the quantum noise in the blue-violet laser is eight times higher than that in the near infrared lasers in terms of RIN at the same output power [11]. To the best of the authors' knowledge, no theoretical investigations on the noise issue in these short-wavelength laser candidates have been reported.

The dynamics and noise of semiconductor lasers are described by a set of stochastic rate equations that describe the mechanisms of time evolution of the adding/dropping of photons and charge carriers [21]. The intensity noise is characterized in terms of the spectral properties of RIN. Analysis of noise and dynamics in semiconductor lasers is dependent on the form of the optical gain in the rate equations and more realistic model should include the effect of the gain suppression [22]. Abdullah [23] indicated that the gain suppression has an important effect on the dynamics of InGaN-based laser diodes, because these lasers may operate with high power where the gain suppression is pronounced. Ahmed and Yamada [24] showed that, in the limit of high power, nonlinear gain is inaccurately described by the commonly used third-order gain expression, which corresponds to partial homogeneous broadening of gain [25]. Instead, the nonlinear gain expression should include higher-order terms from the infinite gain expansion in terms of the electric field intensity to account for the case of high power emission [23].

In this paper, we introduce modeling and simulation on the spectral properties of RIN in the blue-violet InGaN laser diode. The studies are based on appropriate modeling of the rate equations and intensive computer simulations of the laser dynamics and noise. We enhance the novelty of the present work by comparing the simulated results with the previous publications such as in [11]. In addition, we compare the noise results of 410 nm InGaN lasers with those of 830 nm AlGaAs lasers. Also, we study the influence of gain suppression on the noise properties and compare the noise results when using the expressions of inhomogeneous gain broadening. We show that RIN is suppressed remarkably with the increase in the injection current in the regime near the threshold level and that the RIN level in the InGaN laser is nearly 9 dB higher than that of the AlGaAs laser. The noise results are compared with the experimental results in [11]. The increase in the gain suppression was found to suppress the quantum noise within 1 dB/Hz. Finally, we point out that the case of in homogeneously broadened gain overestimates the RIN level by about 4 dB in the limit of high power emission.

THEORETICAL MODEL

The noise properties of the InGaN laser are described by solving the following rate equations of the photon number $S(t)$ and injected carriers $N(t)$ in the active region at a given current I:

$$\frac{dS}{dt} = (G - G_{th})\,S + \frac{a\xi}{V}N + F_S(t)$$

$$\frac{dN}{dt} = \frac{I}{e} - \frac{N}{\tau_s} - G_L S + F_N(t),$$

(1)

Where G is the optical gain and is composed of linear term G_L and nonlinear term G_{NL} [25],

$$G = G_L - G_{NL} S.$$

(2)

The linear term G_L is a linear function of N and is characterized by the gain slope a and the injected carrier number at transparency N_g as [25]

$$G_L = \frac{a\xi}{V}\left(N - N_g\right),$$

(3)

Whereas the nonlinear gain G_{NL} is given in terms of the injected carrier number N as [25]

$$G_{NL} = B_c\left(N - N_s\right),$$

(4)

Where B_c and N_s are characteristic parameters of the nonlinear gain The influence of the nonlinear gain is to suppress the optical gain under the threshold gain level G_{th} due to the increase in the photon number S when the laser operates above threshold [24]. In (3), ξ is the confinement factor of the optical field in the active layer whose volume is V. The other parameters in (1) include e as the electron charge and τs as the electron lifetime due to the spontaneous emission. The last terms $F_s(t)$ and $F_s(t)$ are Langevin noise sources with zero mean values and are added to the equations to account for intrinsic fluctuations in $S(t)$

and $N(t)$, respectively [9]. These noise sources are assumed to have Gaussian probability distributions and to be δ-correlated processes [9]. The frequency content of intensity fluctuations is measured in terms of RIN, which is calculated from the fluctuations $\delta S(t) = S(t) - \bar{S}$ in $S(t)$, where $\bar{S}$ is the time-average value of $S(t)$. Over a finite time T, RIN is given as [14]

$$\text{RIN} = \frac{1}{\bar{S}^2} \left\{ \frac{1}{T} \left| \int_0^T \delta S(t) e^{-j2\pi f\tau} d\tau \right|^2 \right\},$$

(5)

Where f is the noise frequency The noise performance of the laser is evaluated also in terms of the average value of the RIN components at frequencies lower than 100MHz, LF-RIN. The power $P(t)$ emitted from the front facet is determined from the photon number $S(t)$ as [9]

$$P(t) = \frac{hc^2}{2n_D L_D \lambda} \frac{\left(1 - R_f\right) \ln\left(1/R_f R_b\right)}{\left(1 - \sqrt{R_f R_b}\right)\left(1 - \sqrt{R_f R_b}\right)} S(t),$$

(6)

Where λ is the emission wavelength, h is Planck's constant, and R_f and R_b are the power reflectivities at the front and the back facets, respectively.

NUMERICAL CALCULATIONS

Rates (1) are solved by the 4th order Runge-Kutta method using a time integration of $\Delta T = 10$ ps. At each integration instant, the noise sources $F_S(t)$ and $F_N(t)$ are generated following the technique developed in [14] using two uniformly distributed random numbers generated by the computer. RIN of the total output is calculated from the fast Fourier transforms (FFT) of the time fluctuations $\delta S(t_i)$ via (5) as follows:

$$\text{RIN} = \frac{1}{\bar{S}^2} \frac{\Delta t^2}{T} \left| \text{FFT}\left[\delta S(t_i)\right]\right|^2.$$

(7)

In the calculations, we assume 410 nm multiple quantum well (MQW) InGaN single mode laser with the parameters listed in Table

1. The active region is assumed to contain three quantum wells (QWs) with well thickness, barrier thickness, and stripe width of 5 nm, 10 nm, and 5 µm, respectively. The optical confinement factor in each quantum well layer is 0.0342.

Table 1: Definition and numerical values of the InGaN laser parameters

Symbol	Definition	Value
λ	Wavelength	410 nm
n_D	Refractive index of active layer	2.6
L_D	Active layer length	300 μm
a	Differential gain coefficient	1.85.× 10^{-12}m^2
B_c	Nonlinear gain coefficient	2.7.×10^{-5}s^{-1}
N_g	Carrier.number.at.transparency	2.52.×.10^8
N_s	Carrier number characterizing	2.01.×.10^8
α	Linewidth enhancement factor	2
τ_s	Spontaneous emission lifetime	2 ns
R_f	Front facet reflectivity	0.4
R_b	Back facet reflectivity	0.7
G_{th}	Threshold gain	2.2.×10^{11}s^{-1}

RESULTS AND DISCUSSION

RIN in InGaN Laser Diodes

Figure 1 plots the spectral characteristics of RIN at different injection levels: near above threshold (I = 1.01 and 1.1I_{th}), above threshold (I = 1.5I_{th}), and far above threshold (I = 3.0I_{th}). The RIN spectra are flat (white noise) in the low-frequency regime due to the small amplitude of the intensityfluctuations and the large signal-tonoise ratio. In the high-frequency regime, the spectra exhibit the well-known carrier-photon resonance peak around the relaxation frequency f_r. The figure shows suppression of RIN with the increase in the injection current

I due to the improvement in the degree of laser coherency [9]. This increase in the injection level is associated also with an increase in the relaxation oscillation peak; $f_r = 330\text{MHz}$ when $I = 1.01I_{th}$ and $f_r = 5.1\text{GHz}$ when $I = 3.01I_{th}$.

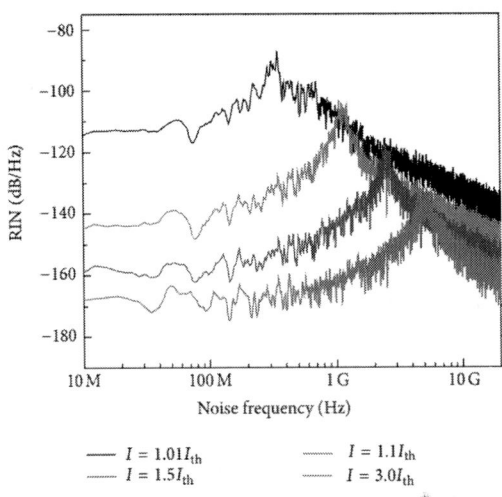

Figure 1: Spectra of RIN as functions of I/I_{th} for InGaN lasers.

It is of practical interest to study variation of the low frequency level of RIN and LF-RIN with the increase in the injection current I. Figure 2 plots such a variation, showing that LF-RIN increases with the increase in I around the threshold current I_{th}. The further increase in I results in a drop of LF-RIN to lower orders of magnitude up to $I \sim 1.6I_{th}$ (LF-RIN decreases from the peak of −91.8 dB/Hz when $I = I_{th}$ to −157 dB/Hz when $I = 1.6I_{th}$). When I increases beyond $1.6I_{th}$, the decrease in LF-RIN with the increase in I becomes as small as ~0.5 dB/Hz. This suppression of RIN is associated with a decrease in the amplitude of intensity fluctuations in the laser signal and improvement in the signal to noise ratio [9]. We plot also in Figure 2 the experimental results on the quantum noise of InGaN laser reported by Matsuoka et al. [11]. The figure shows that the simulated noise results fit qualitatively the experimental data in [11]. The differences in the noise level between the simulated and experimental results could be a mode competition effect in the measured laser, which may change the noise level of the modeled single mode laser [26].

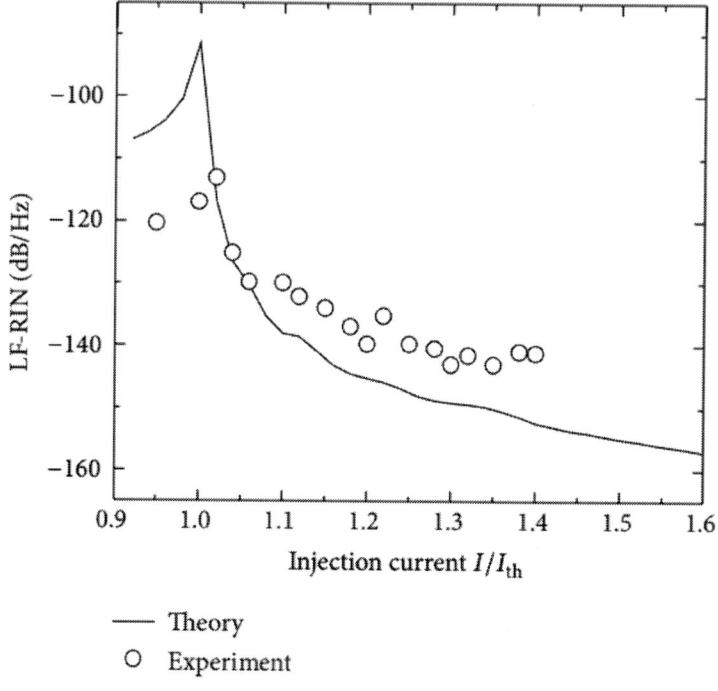

Figure 2: Variation of LF-RIN with the injection current I/I_{th} of InGaN lasers (solid). The experimental results in [11] are shown (circles) for comparison.

Comparison of RIN in InGaN Lasers with InGaN and AlGaAs Lasers

In this subsection, we compare the lasing characteristics of the 410 nm GaInN laser with those of 830 nm AlGaAs laser diode. For AlGaAs laser, we consider the following parametric values: $a = 2.7 \times 10^{-12}$ m², $n_D = 3.59$, $N_g = 1.89 \times 10^8$, $N_s = 1.63 \times 10^8$, $B_c = 3.95 \times 10^{-5}$ s⁻¹ and $\tau_s = 2.7$ ns. Figure 3 compares the L-I characteristics of the two laser diodes showing that the laser operation is linear over the relevant range of injection current I. The threshold current I_{th}, which is the intercept of the L-I line with the current axis, is shown to increase from 14.2mA in the AlGaAs laser to 28.2mA in the InGaN laser.

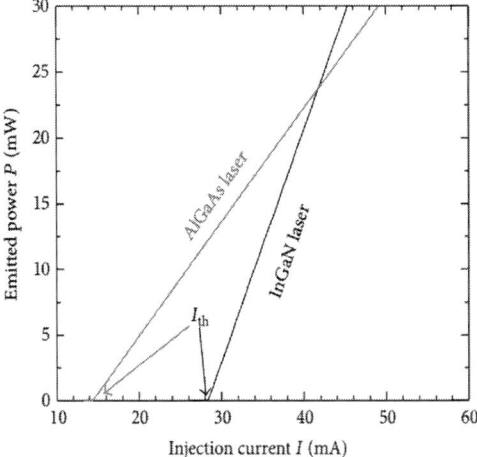

Figure 3: The *L-I* characteristics of the InGaN and AlGaAs laser diodes.

Figure 4 plots variation of the LF-RIN level with the emitted power for the two laser diodes. The figure shows the effect of noise suppression with the increase in the lasing power *P*, with the suppression being stronger in the regime of the near above threshold. The figure shows that LF-RIN in the InGaN laser is nearly 9 dB higher than that of the AlGaAs laser. This higher level of the LF-RIN in the InGaN laser is due to the inverse proportionality of RIN on the third power of the emission wavelength λ [10]:

$$\text{RIN} \propto \frac{1}{\lambda^3 P^3}.$$

(8)

Therefore, this relation predicts that the quantum noise in the 410 nm InGaN laser is nearly 8.3 dB higher than that in the AlGaAs laser. This result fits the experimental finding by Matsuoka et al. [11] on 410 nm InGaN and 830 nm AlGaAs lasers with almost identical structure design and operation characteristics [11]. It is worth noting that the corresponding variation of the quantum noise level with the current ratio I/I_{th} reveals much smaller differences between both lasers, which agree also with the experimental results in [11].

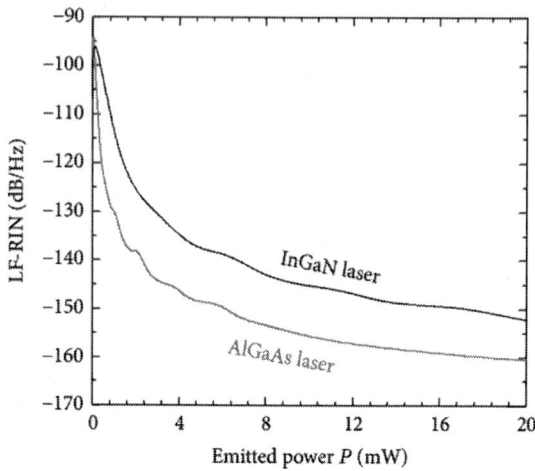

Figure 4: Variation of LF-RIN as a function of the emitted power P.

Figure 5 compares the simulated spectrum of RIN of both lasers when P=5 mW. As shown in the figure, the InGaN laser reveals RIN spectra higher than those of the AlGaAs laser. On the other hand, the position of the enhanced resonance peak of the InGaN laser occurs at a relaxation frequency lower than that of the AlGaAs laser.

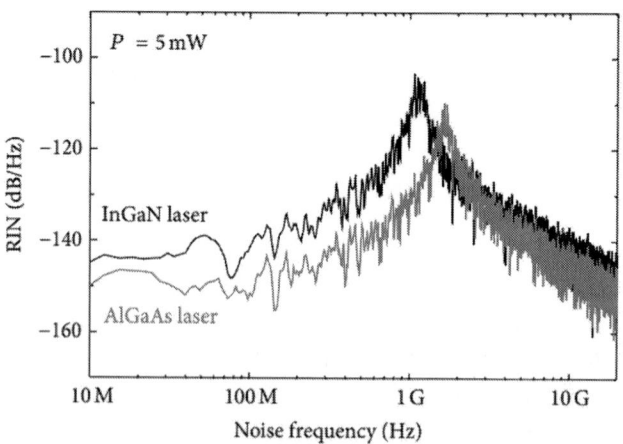

Figure 5: The spectra of RIN of InGaN and AlGaAs lasers when P=5 mW.

Influence of Gain Saturation on RIN

The gain suppression is a critical property of the lasing medium and controls most of the dynamics and operation characteristics of semiconductor lasers. Influence of the nonlinear gain suppression G_{NL} on the quantum noise of the InGaN laser is illustrated in Figure 6 when $I = 5I_{th}$. The gain suppression is varied by varying B_c relative to the set value in Table 1, B_{c0}. Thefigure shows that the increase in G_{NL} results in a decrease in the LF-RIN level within 1 dB/Hz. The variation G_{NL} was found not to affect the position of the resonance peak of the RIN spectrum (i.e., the relaxation frequency). This suppression in the amplitude of intensity fluctuations agrees with the prediction by Abdulrhmann et al. [27] in InGaAsP lasers.

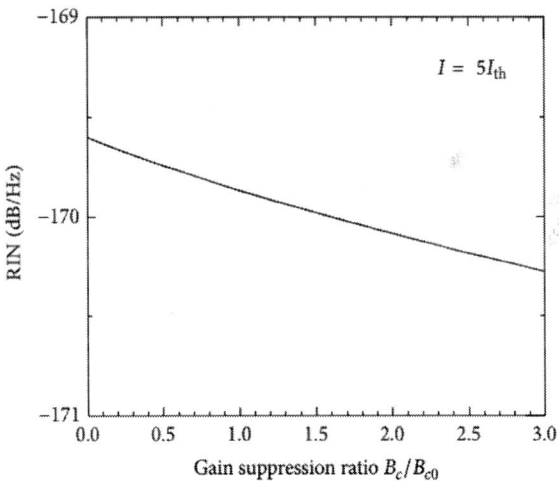

Figure 6: Influence of gain suppression factor on the LF-RIN level when $I = 5I_{th}$.

The spectral gain suppression is a nonlinear effect. In its most exact description, the optical gain is expressed as an infinite perturbation expansion of the field intensity [24]. Yamada and Suematsu [25] showed that within the normal operation of semiconductor lasers the gain suppression has partial homogeneous spectral broadening. Therefore, the gain of semiconductor lasers is usually described by the third-order perturbation form in (2) [25]. In semiconductor lasers with high power emission, the gain suppression becomes inhomogeneously broadened [25] and the infinite gain expansion in [23] is reduced to the following form developed by Agrawal [28]:

$$G = \frac{G_L}{\sqrt{1 + \varepsilon S}} \quad \text{where } \varepsilon = B_c \frac{V}{a\xi}.$$

(9)

Here, we examine the validity of this gain suppression to model the noise properties of the present InGaN laser. Replacing form (2) of the partial homogeneous gain suppression with form (9) of the inhomogeneous gain suppression was found not to change the *L-I* characteristics. In Figure 7, we compare the obtained noise properties of the InGaN laser by those obtained when applying the form in (9). As shown in the figure, the inhomogeneous gain broadening little overestimates the RIN level in the regime of high injection currents; the difference is just within 4 dB.

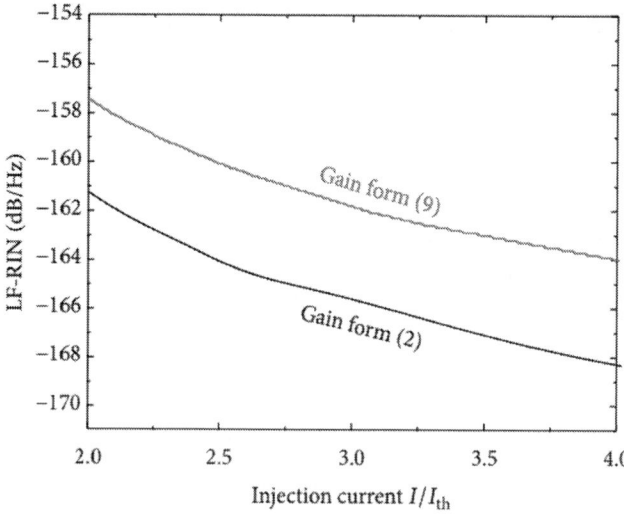

Figure 7: Dependence of LF-RIN on the injection current I/I_{th} of InGaN lasers under partial homogeneous gain broadening (form (2)) and inhomogeneous gain broadening (form (9)).

CONCLUSIONS

We modeled and simulated the noise properties of the blue-violet InGaN laser diodes. We compared the RIN results with those of the AlGaAs lasers since both lasers are the most representative light sources in the high-density disc systems. Also, we examined the influence of the gain suppression on RIN. The results showed that the RIN spectra are flat in the low-frequency regime. RIN is suppressed remarkably with the increase in the injection current up to $I \sim 1.6I_{th}$, beyond which the decrease in LF-RIN becomes within 0.5 dB/Hz This increase in the injection level is associated also with an increase in the relaxation oscillation peak towards higher frequencies. The simulated noise results are in good agreement with the experimental data reported in literature. The RIN level in the InGaN laser is nearly 9 dB higher than that of the AlGaAs laser. A big contributor to this higher RIN is the inverse proportionality of RIN on the third power of the emission wavelength. The increase in the gain suppression was found to decrease the quantum noise within 1 dB/Hz. Describing the gain suppression by the form of inhomogeneous gain broadening was found to induce 4 dB overestimation of the RIN level.

ACKNOWLEDGMENT

This work was funded by the Deanship of Scientific Research (DSR), King Abdulaziz University, Jeddah, under Grant no. (130 - 474 - D1435) The author, therefore, acknowledges with thanks DSR technical and financial support.

REFERENCES

1.   J. Piprek, Nitride Semiconductor Devices, Wiley-VCH, Weinheim, Germany, 2007.

2.   S. Nakamura, S. Pearton, and G. Fasol, The Blue Laser Diode, Springer, Berlin, Germany, 2nd edition, 2000.

3.   S. Nakamura, M. Senoh, S. Nagahama et al., "Optical gain and carrier lifetime of InGaN multi-quantum well structure laser

diodes," Applied Physics Letters, vol. 69, no. 11, pp. 1568–1570, 1996.

4.   T. Takeuchi, H. Takeuchi, S. Sota, H. Sakai, H. Amano, and I. Akasaki, "Optical properties of strained AlGaN and GaInN on GaN," Japanese Journal of Applied Physics, vol. 36, no. 2, pp. L177–L179, 1997.

5.   T. Kobayashi, F. Nakamura, K. Naganuma et al., "Room-temperature continuous-wave operation of GaInN/GaN multiquantum well laser diode," Electronics Letters, vol. 34, no. 15, pp. 1494–1495, 1998.

6.   A. Kuramata, S. Kubota, R. Soejima, K. Domen, K. Horino, and T. Tanahashi, "Room-temperature continuous wave operation of InGaN laser diodes with vertical conducting structure on SiC substrate,"Japanese Journal of Applied Physics, vol. 37, no. 11, pp. L1373–L1375, 1998.

7.   M. Kuramoto, C. Sasaoka, Y. Hisanaga et al., "Room-temperature continuous-wave operation of InGaN multi-quantum-well laser diodes grown on an n-GaN substrate with a backside n-contact," Japanese Journal of Applied Physics, vol. 38, part 2, pp. L184–L186, 1999.

8.   M. Kneissl, D. P. Bour, C. G. van de Walle et al., "Room-temperature continuous wave operation of InGaN," Moldavian Journal of the Physical Sciences, vol. 5, pp. 153–170, 2006.

9.   M. Ahmed, M. Yamada, and M. Saito, "Numerical modeling of intensity and phase noise in semiconductor lasers," IEEE Journal of Quantum Electronics, vol. 37, no. 12, pp. 1600–1610, 2001.

10.   M. Yamada, "Variation of intensity noise and frequency noise with the spontaneous emission factor in semiconductor lasers," IEEE Journal of Quantum Electronics, vol. 30, no. 7, pp. 1511–1519, 1994.

11.   K. Matsuoka, K. Saeki, E. Teraoka, M. Yamada, and Y. Kuwamura, "Quantum noise and feed-back noise in blue-violet InGaN semiconductor lasers," IEICE Transactions on Electronics, vol. E89-C, no. 3, pp. 437–439, 2006.

12.   M. Ahmed, "Numerical characterization of intensity and frequency fluctuations associated with mode hopping and single-mode jittering in semiconductor lasers," Physica D, vol. 176, no. 3-4, pp. 212–236, 2003.

13. M. Ahmed, "Spectral lineshape and noise of semiconductor lasers under analog intensity modulation,"Journal of Physics D: Applied Physics, vol. 41, Article ID 175104, 10 pages, 2008.

14. S. Abdulrhmann, M. Ahmed, and M. Yamada, "A new model of analysis of semiconductor laser dynamics under strong optical feedback in fiber communication systems," in The International Society for Optical Engineering: Physics and Simulation of Optoelectronic Devices XI, vol. 4986 of Proceedings of SPIE, pp. 490–501, January 2003.

15. A. W. Smith and J. A. Armstrong, "Intensity noise in multimode GaAs laser emission," IBM Journal of Research and Development, vol. 10, pp. 225–232, 1966.

16. C. H. Henry, "Phase noise in semiconductor lasers," Journal of Lightwave Technology, vol. 4, no. 3, pp. 298–311, 1986.

17. G. P. Agrawal, N. A. Olsson, and N. K. Dutta, "Effect of fiber-far-end reflections on intensity and phase noise in InGaAsP semiconductor lasers," Applied Physics Letters, vol. 45, no. 6, pp. 597–599, 1984.

18. G. P. Agrawal, "Noise in semiconductor lasers and its impact on optical communication systems," inLaser Noise, Proceedings of SPIE, pp. 224–235, November 1990.

19. G. P. Agrawal, "Effect of nonlinear gain on intensity noise in single-mode semiconductor lasers,"Electronics Letters, vol. 27, no. 3, pp. 232–234, 1991.

20. M. Yamada, W. Ishimori, H. Sakaguchi, and M. Ahmed, "Time-dependent measurement of the mode competition phenomena among longitudinal modes in long-wavelength Lasers," IEEE Journal of Quantum Electronics, vol. 39, no. 12, pp. 1548–1554, 2003.

21. M. Ahmed, "Numerical approach to field fluctuations and spectral lineshape in InGaAsP laser diodes,"International Journal of Numerical Modelling: Electronic Networks, Devices and Fields, vol. 17, no. 2, pp. 147–163, 2004.

22. C. Masoller, C. Cabeza, and A. S. Schifino, "Effect of the nonlinear gain in the visibility of a semiconductor laser with incoherent feedback in the coherence collapsed regime," IEEE Journal of Quantum Electronics, vol. 31, pp. 1022–1028, 1995.

23. R. Abdullah, "The influence of gain suppression on dynamic characteristics of violet InGaN laser diodes," Optik, vol. 125, pp. 580–582, 2014.

24. M. Ahmed and M. Yamada, "An infinite order perturbation approach to gain calculation in injection semiconductor lasers," Journal of Applied Physics, vol. 84, no. 6, pp. 3004–3015, 1998.

25. M. Yamada and Y. Suematsu, "Analysis of gain suppression in undoped injection lasers," Journal of Applied Physics, vol. 52, no. 4, pp. 2653–2664, 1981.

26. M. Ahmed, "Numerical characterization of intensity and frequency fluctuations associated with mode hopping and single-mode jittering in semiconductor lasers," Physica D: Nonlinear Phenomena, vol. 176, no. 3-4, pp. 212–236, 2003.

27. S. Abdulrhmann, M. Ahmed, and M. Yamada, "Influence of nonlinear gain and nonradiative recombination on the quantum noise in InGaAsP semiconductor lasers," Optical Review, vol. 9, no. 6, pp. 260–268, 2002.

28. G. P. Agrawal, "Effect of gain nonlinearities on period doubling and chaos in directly modulated semiconductor lasers," Applied Physics Letters, vol. 49, no. 16, pp. 1013–1015, 1986.

Novel Optoelectronic Devices Based on Single Semiconductor Nanowires (Nanobelts)

Yu Ye[1], Lun Dai[1], Lin Gan[2], Hu Meng[1], Yu Dai[1],
Xuefeng Guo[2], and Guogang Qin[1]

[1]State Key Lab for Mesoscopic Physics and School of Physics, Peking University, Beijing, 100871, China
[2]College of Chemistry and Molecular Engineering, Peking University, Beijing, 100871, China

ABSTRACT

Semiconductor nanowires (NWs) or nanobelts (NBs) have attracted more and more attention due to their potential application in novel optoelectronic devices. In this review, we present our recent work on novel NB photodetectors, where a three-terminal metal–semiconductor field-effect transistor (MESFET) device structure was exploited. In contrast to the common two-terminal NB (NW) photodetectors, the MESFET-based photodetector can make a balance among overall performance parameters, which is desired for practical device applications. We also present our recent work on graphene nanoribbon/semiconductor NW (SNW) heterojunction light-emitting diodes (LEDs). Herein, by taking advantage of both graphene and SNWs, we have fabricated, for the first time, the graphene-based nano-LEDs. This achievement opens a new avenue for developing graphene-based nano-electroluminescence devices. Moreover, the novel graphene/SNW hybrid devices can also find use in other applications, such as high-sensitivity sensor and transparent flexible devices in the future.

REVIEW

Introduction

Semiconductor single-crystalline nanowires (NWs) or nanobelts (NBs) can be grown on lattice mismatched substrates and constructed into devices with the bottom-up method on basically any substrates [1]. Hence, compared to the conventional ones, semiconductor NW- or NB-based devices have the advantage of versatility in both the material and the device structure. So far, various semiconductor NW- or NB-based nanodevices have been emerging continuously [2-4]. Developing novel high-performance nano-optoelectronic devices is not only important in diverse device applications, but also has significant meaning in exploring and realizing optoelectronic integration.

In this review, we present our research work on two types of novel optoelectronic devices based on semiconductor NWs (NBs). One is semiconductor NB metal–semiconductor field-effect transistor (MESFET)-based photodetectors [5]. In contrast to the common two-terminal single semiconductor NB (NW) photodetectors, the three-terminal NB MESFET-based photodetector can make a balance among overall performance parameters, which is desired for practical device applications. The other is novel multicolor light-emitting diodes (LEDs) based on graphene nanoribbon (GNR)/semiconductor nanowire (SNW) heterojunctions [6]. Herein, ZnO, CdS, and CdSe NWs were employed for demonstration. At forward biases, the GNR/SNW heterojunction LEDs emitted light from ultraviolet (380 nm) to red (705 nm), which were determined by the bandgaps of the involved SNWs. This work opens a new avenue for developing diverse graphene-based optoelectronic devices [7]. These two works may help to promote nano-optoelectronic integration in the future.

Single CdS NB MESFET Photodetector

Photodetectors, which convert light to electric signals, are essential elements in high-resolution imaging techniques and light-wave communication, as well as in future memory storage [8]. Single NB (NW) photodetectors may find applications as binary switches, light-wave communications, and optoelectronic circuits. So far, most of the reported single NB (NW) photodetectors are two-terminal devices [3, 8-17]. We summarize the key parameters of the three-terminal MESFET CdS NB photodetector and the reported two-terminal CdS NB (NW) photodetectors in Table 1. In general, for the two-terminal NB (NW) photodetectors, there exists a trade-off among the performance parameters, such as current responsivity (R_λ), photoresponse ratio (I_{light}/I_{dark}), and photoresponse time (rise and fall times). For example, Golberg et al. reported ohmic contact-based single CdS NB photodetectors with ultrahigh R_λ (approximately 7.3×10^4 A/W) and fast response time (approximately $20\,\mu s$ of both rise and fall times); however, the I_{light}/I_{dark} was quite low (approximately 6) [15]. Wang et al. reported Schottky contact-based NW photodetectors with a higher I_{light}/I_{dark} (approximately183); however, the response time was not satisfying (approximately $320\,ms$ of fall time) [17]. Compared to the reported two-terminal NB (NW) photodetectors, the MESFET-based photodetector can make a balance among these key parameters and have an overall improvement in the device performance.

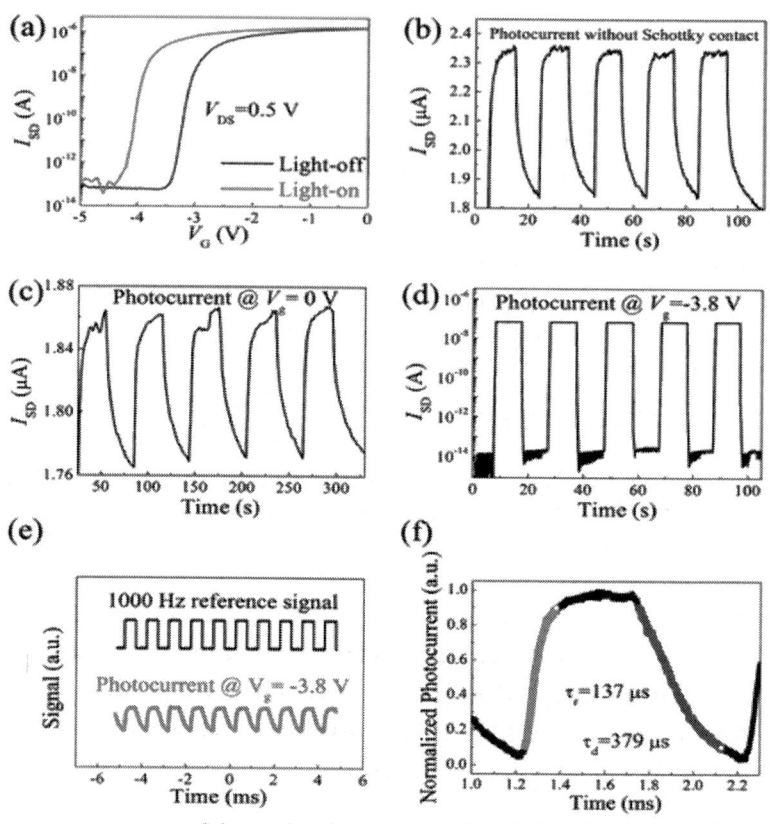

Figure 2: The typical light response properties of the single CdS NB MESFET-based photodetectors. (a) The transfer characteristics of a CdS NB MESFET-based photodetector measured in the dark (black line) and under illumination (red line). (b) On/off photocurrent response of the CdS NB without Schottky contact as a function of time. (c) On/off photocurrent response of the CdS NB MESFET-based photodetector with $V_G = 0$ V as a function of time on a linear scale. (d) On/off photocurrent response of the CdS NB MESFET-based photodetector with $V_G = -3.8$ V as a function of time on an exponential scale. (e) A transient response of the CdS NB MESFET-based photodetector ($V_G = -3.8$ V, $V_{DS} = 0.5$ V) along with a reference signal of the chopped light with a frequency of 1,000 Hz. (f) A close-up of the result shown in (e).

Figure 2d shows the on/off photocurrent response of the CdS NB MESFET-based photodetector measured at $V_G = -3.8$ V, which is the threshold voltage of the MESFET under light illumination. We can see that the average dark current and photocurrent are about 26 fA and

70 nA, respectively, resulting in a I_{light}/I_{dark} as high as approximately 2.7×10^6. To the best of our knowledge, this is so far among the highest reported values for single NB (NW) photodetectors [3, 8, 9-17]. In addition, the photoresponse processes (both rise and decay processes) are quite fast, which have exceeded the detection limit (0.3 s) of the measurement apparatus (Keithley 4200, Cleveland, OH, USA).

The R_λ, defined as the photocurrent generated per unit power of incident light on the effective illuminated area of a photoconductor, and the external quantum efficiency (EQE), defined as the number of electrons detected per incident photon, are two critical parameters for photodetectors. The R_λ and EQE can be calculated with equations

$R_\lambda = \dfrac{\Delta I}{P_\lambda S}$ and $EQE = \dfrac{hcR_\lambda}{e\lambda}$ [11], respectively. Here, ΔI is the difference between the photocurrent and the dark current, P_λ is the light power density, S is the effective illuminated area, h is Planck›s constant, c is the velocity of light, e is the electronic charge, and λ is the light wavelength. Using $\Delta I = 7.0 \times 10^{-8}$ A (measured from Figure 2d), $P_\lambda = 5.3$ mW/cm^2 S = 500 nm $\times$ 13 μm (measured from the inset of Figure 1a), $\lambda = 488$ nm, the R_λ and EQE of the CdS NB MESFET photodetector can be estimated to be approximately 2.0×10^2 A/W and 5.2×10^2, respectively.

In order to further investigate the detailed photoresponse times of the single CdS NB MESFET photodetector, we employed a 200-MHz digital oscilloscope (Tektronix DPO2024, Beaverton, OR, USA) with a 10-MΩ impedance and an optical chopper working at a frequency of 1,000 Hz, as shown in Figure 2e. From the close-up of the measured result shown in Figure 2f, the rise timer, defined as the time needed for the photocurrent to increase from 10 % i_{peak} to 90 % i_{peak}, is 137 μs and the decay timed, defined analogously, is 379 μs.

We attribute the overall high performance of our CdS NB MESFET-based photodetectors to the unique advantage of the MESFET structure. Compared to two-terminal photodetectors, there are two main advantages of the MESFET-based photodetectors. First, it has a much lower dark current because the applied negative gate voltage (in our case, the threshold voltage under illumination) helps to deplete the channel carriers. Second, this gate depletion effect will also cause a fast current recovery when the light is turned off. Consequently, the decay tail, which is normally observed in a two-terminal photodetectors, is suppressed in the MESFET-based photodetectors.

Multicolor GNR/SNW Heterojunction LEDs

By taking advantage of both graphene and SNWs, we have fabricated, for the first time, the graphene-based nano-LED [6]. This achievement opens a new avenue for developing graphene-based nano-electroluminescence devices. Moreover, the novel graphene/SNW hybrid devices can also find use in other applications, such as high-sensitivity sensor and transparent flexible devices in the future.

Both the n-type NWs [20-22] and the graphene [23] used in this work were synthesized via the CVD method. Before device fabrication, the graphenes were transferred by the stamp method with the help of polymethyl methacrylate [24] to Si/300-nm SiO_2 substrates for Raman and electrical property characterizations, to quartz substrates for transparency characterization, and transferred to carbon-coated grids for high-resolution transmission electron microscopy (HRTEM) characterization (Tecnai F30, FEI, Eindhoven, The Netherlands). Their electrical properties were measured by a Hall effect measurement system (Accent HL5500, York, England).

The HRTEM, Raman, and transparency characterization results for the as-synthesized graphenes (Figure 3) demonstrate that the graphenes have high quality, monolayer, and high transparency. The typical sheet resistance, hole concentration, and hole mobility of the graphenes are about 345 Ω/sq, 1.84×10^{14} cm^{-2}, and 98.6 cm^2/V·s, respectively.

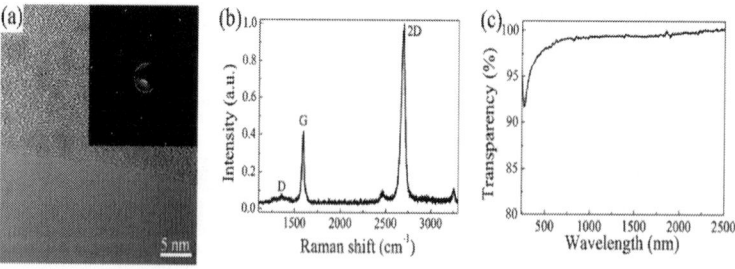

Figure 3: Properties of as-synthesized graphene. (a) Typical HRTEM image of an as-synthesized graphene, indicating the formation of monolayer graphene.

Inset: selected area electron diffraction pattern of the graphene. (b) Raman spectrum of an as-synthesized graphene on a Si/300-nm SiO$_2$ substrate. (c) The transparency spectrum of the graphene on a quartz substrate.

The fabrication processes of a GNR/SNW heterojunction LED are shown in Figure 4. Figure 5a shows an FESEM image of an as-fabricated GNR/CdS NW heterojunction LED. The IV curve (Figure 5b) of the LED shows an excellent rectification characteristic. An on/off current ratio of approximately 3.4×10^7 can be obtained when the voltage changes from +1.5 to −1.5 V. The turn-on voltage is around 1.1 V. In view of the high conductivity and near-zero bandgap characteristics of the GNR [25], the heterojunction structure of the GNR/CdS NW can be considered approximately as a metal–semiconductor contact of the Schottky model [26]. We can deduce that the diode ideality factor n = 1.58. Note that the GNR/ZnO NW and GNR/CdSe NW heterojunctions show similar rectification characteristics as described above, with the turn-on voltages to be about 0.7 and 1.2 V, respectively.

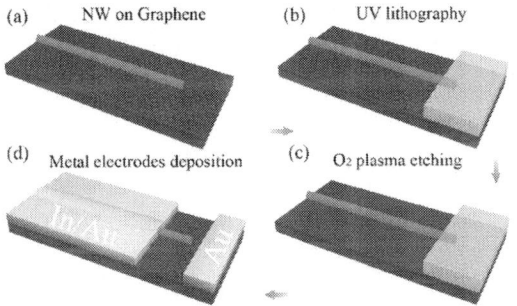

Figure 4: Schematic illustration of the fabrication processes of a GNR/SNW heterojunction LED. (a) The as-synthesized large-scale graphene was transferred to a Si/SiO$_2$ substrate. After that, SNW suspension was dropped on the graphene. (b) A photoresist pad was patterned to cover one end of a SNW by UV lithography and development processes. (c) Oxygen plasma etching was used to remove the exposed graphene. After that, the GNR formed under the SNW. (d) After removing the photoresist, In/Au and Au ohmic contact electrodes to SNW and graphene pad were defined, respectively. It is worth noting that because an undercut was formed during the oxygen plasma etching process (Figure 4c) [6, 27], the In/Au electrode on the SNW will not contact with the GNR beneath.

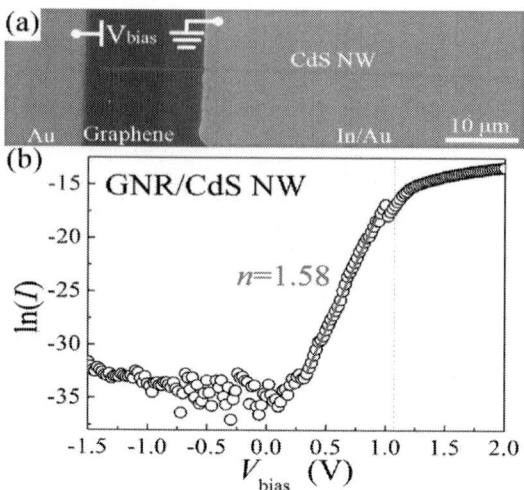

Figure 5: FESEM image and room-temperature I-V characteristic of as-fabricated GNR/CdS NW heterojunction LED. (a) FESEM image of an as-fabricated GNR/CdS NW heterojunction LED. (b) Room-temperature I-V characteristic of the LED in (a) on a semilog scale. The red straight line shows the fitting result of the I-V curve by the equation

$$\ln(I) = \frac{qV}{nkT} \ln(I_0).$$

Figure 6a,b,c shows the electroluminescence (EL) images (Olympus BX51M, Shinjuku-ku, Japan) of the GNR/SNW (ZnO, CdS, CdSe, respectively) heterojunction LEDs at a forward bias of 5 V. Except for the ZnO NW case (where the emitting light is invisible ultraviolet light) in Figure 6a, strong emitting light spots can be seen clearly with naked eyes at the exposed ends of the NWs. For the CdS NW case (Figure 6b), we can see another glaring light spot on the NW. This may be due to the scattering from the defect or adhered particle on the CdS NW [28]. Figure 6d,e,f shows the room-temperature EL spectra measured at various forward biases for the GNR/SNW heterojunction LEDs, where the SNWs are ZnO, CdS, and CdSe NWs, respectively. For all the LEDs, EL intensities increase with the forward biases. The peak wavelength of each EL spectrum (380, 513, and 705 nm, respectively, from (d) to (f)) coincides with the band-edge emission of the SNW involved. This indicates that the radiative recombination of electrons and holes occurs in the SNWs.

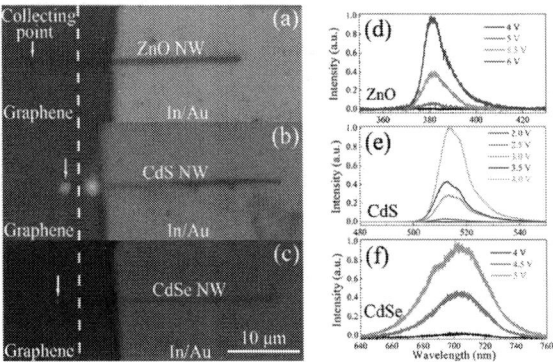

Figure 6: Optical images and room-temperature EL spectra of the GNR/SNW heterojunction LEDs. (a, b, c) The optical images of the GNR/SNW (ZnO, CdS, CdSe, respectively) heterojunction LEDs at a forward bias of 5 V. Dashed lines were used to demarcate the graphenes from the substrates. White arrows: the light collecting points during the EL measurements. (d, e, f) Room-temperature EL spectra for GNR/SNW (ZnO, CdS, CdSe, respectively) heterojunction LEDs at various forward biases.

We can qualitatively understand the mechanism of the light emitting for the GNR/SNW heterojunction LEDs by studying the energy band diagrams. Figure 7a shows the thermal equilibrium energy band diagram of a graphene/n-type semiconductor structure, where the work function of graphene is Φ, and the electron affinity of the semiconductor is χ. E_g and E_F correspond, respectively, to the bandgap and the Fermi level of the semiconductor. It is worth noting that because the graphene used in this work has a very high conductivity and can be taken as a metal, the graphene/SNW heterostructure herein can be taken as a kind of Schottky junction. At the thermal equilibrium contacting state, the energy band of the semiconductor will bend upward at the graphene/semiconductor interface due to the difference between their work functions, and the Fermi levels at the two sides are brought into coincidence. Under a forward bias (i.e., a positive bias on graphene), the built-in potential is lowered. Therefore, more electrons will flow from n-type semiconductor to graphene, and simultaneously, more holes will flow from graphene to n-type semiconductor. Herein, the injected holes have a higher radiative recombination with the electrons in the SNW (the direct bandgap semiconductor). Accordingly, the EL spectra are determined mainly by the band-edge emission of the SNWs.

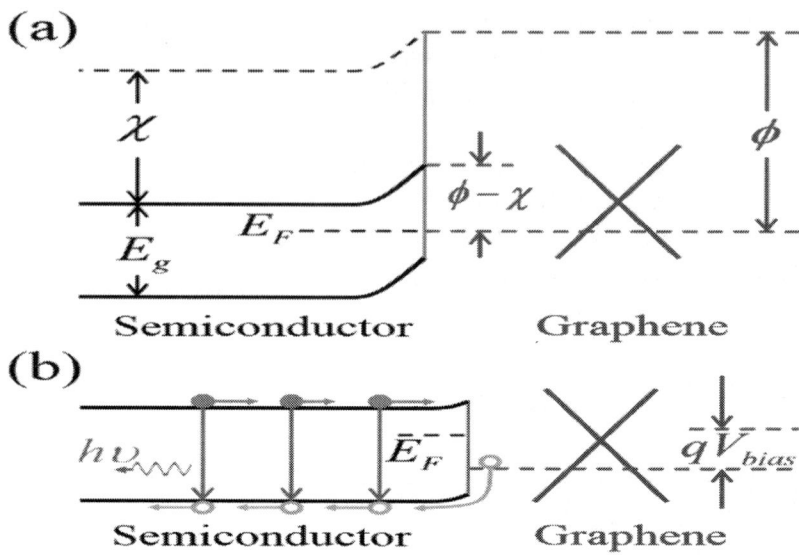

Figure 7: Schematic illustration of the energy band diagrams of a graphene/ semiconductor heterojunction. (a) The thermal equilibrium energy band diagram. (b) The energy band diagram of the heterojunction under a forward bias. Φ: the work function of graphene; χ: the electron affinity of the semiconductor.

It is worth noting that the GNR/NW structure has clear advantage over the conventional Schottky structure. For comparison, we have fabricated various metal/SNW Schottky structures, where the NWs used are identical to those reported in this work. Unfortunately, no EL can be observed in these structures. We attribute this to the well-known luminescence quenching effect caused by the involved metal [29]. Moreover, in our face-to-face contact LED, the active region, where the radiative recombination occurs, is larger and the series resistance is smaller, compared to the crossed NWs or NW/Si pad heterojunction structures [20, 30, 31]. These merits may benefit high-efficiency EL and even electrically driven laser in the future.

CONCLUSIONS

We review two types of novel nano-optoelectronic devices developed in our group recently. One is the photodetector, which converts light to electric signals. Our MESFET-based photodetectors have ultrahigh I_{light}/I_{dark} (approximately 2.7×10^6) and fast response (rise time, approximately 137 µs; decay time, approximately 379 µs) simultaneously. The other is LED, which converts electric power to light. At forward biases, our novel GNR/SNW heterojunction LEDs emitted light with wavelengths varying from ultraviolet (380 nm) to red (705 nm), which were determined by the bandgaps of the involved SNWs. These two types of nano-optoelectronic devices may find diverse applications in future nano-optoelectronic integration.

AUTHORS' CONTRIBUTIONS

YY carried out the device fabrications, participated in the statistical measurements, and drafted the manuscript. LD and GQ participated in the instruction, discussion, and manuscript revision. LG and XG synthesized the graphene. HM participated in the device design. YD synthesized the CdSe NWs. All authors have read and approved the final manuscript.

ACKNOWLEDGMENTS

This work was supported by the National Natural Science Foundation of China (nos. 61125402, 51172004, 11074006, 10874011, 50732001), the National Basic Research Program of China (nos. 2012CB932703, 2007CB613402), and the Fundamental Research Funds for the Central Universities.

REFERENCES

1. Yang PD, Yan RX, Fardy M: Semiconductor nanowire: what's next? Nano Lett 2010, 10:1529.

2. Duan XF, Huang Y, Agarwal R, Lieber CM: Single-nanowire electrically driven lasers. Nature 2003, 421:241.

3. Shen GZ, Chen D: One-dimensional nanostructures for photodetectors. Recent Pat Nanotechnol 2010, 4:20.

4.   Zhai TY, Fang XS, Li L, Bando Y, Golberg D: One-dimensional CdS nanostructures: synthesis, properties, and applications. Nanoscale 2010, 2:168.

5.   Ye Y, Dai L, Wen XN, Wu PC, Pen RM, Qin GG: High-performance single CdS nanobelt metal–semiconductor field-effect transistor-based photodetectors. ACS Appl Mater Interfaces 2010, 2:2724.

6.   Ye Y, Gan L, Dai L, Meng H, Wei F, Dai Y, Shi ZJ, Yu B, Guo XF, Qin GG: Multicolor graphene nanoribbon/semiconductor nanowire heterojunction light-emitting diodes. J Mater Chem 2011, 21:11760.

7.   Bonaccorso F, Sun Z, Hasan T, Ferrari AC: Graphene photonics and optoelectronics. Nat Photon 2010, 4:611.

8.   Kind H, Yan HQ, Messer B, Law M, Yang PD: Nanowire ultraviolet photodetectors and optical switches. Adv Mater 2002, 14:158.

9.   Jie JS, Zhang WJ, Jiang Y, Meng XM, Li YQ, Lee ST: Photoconductive characteristics of single-crystal CdS nanoribbons. Nano Lett 1887, 2006:6.

10.   Gao T, Li QH, Wang TH: CdS nanobelts as photoconductors. Appl Phys Lett 2005, 86:173105.

11.   Zhai TY, Fang XS, Liao MY, Xu XJ, Li L, Liu BD, Koide Y, Ma Y, Yao JN, Bando Y, Golberg D:Fabrication of high-quality In2Se3 nanowire arrays toward high-performance visible-light photodetectors. ACS Nano 2010, 4:1596.

12.   Fang XS, Xiong SL, Zhai TY, Bando Y, Liao MY, Gautam UK, Koide Y, Zhang XG, Qian YT, Golberg D: High-performance blue/ultraviolet-light-sensitive ZnSe-nanobelt photodetectors. Adv Mater 2009, 21:5016.

13.   Fang XS, Bando Y, Liao MY, Gautam UK, Zhi CY, Dierre B, Liu BD, Zhai TY, Sekiguchi T, Koide Y, Golberg D: Single-crystalline ZnS nanobelts as ultraviolet-light sensors. Adv Mater 2034, 2009:21.

14.   Zhai TY, Liu HM, Li HQ, Fang XS, Liao MY, Li L, Zhou HS, Koide Y, Bando Y, Golberg D:Centimeter-long V2O5 nanowires: from synthesis to field-emission, electrochemical, electrical transport, and photoconductive properties. Adv Mater 2010, 22:2547.

15.   Li L, Wu PC, Fang XS, Zhai TY, Dai L, Liao MY, Koide Y, Wang HQ, Bando Y, Golberg D:Single-crystalline CdS nanobelts

for excellent field-emitters and ultrahigh quantum-efficiency photodetectors. Adv Mater 2010, 22:3161.

16. Zhou J, Gu YD, Hu YF, Mai WJ, Yeh PH, Bao G, Sood AK, Polla DL, Wang ZL: Gigantic enhancement in response and reset time of ZnO UV nanosensor by utilizing Schottky contact and surface functionalization. Appl Phys Lett 2009, 94:191103.

17. Wei TY, Huang CT, Hansen BJ, Lin YF, Chen LJ, Lu SY, Wang ZL: Large enhancement in photon detection sensitivity via Schottky-gated CdS nanowire nanosensors. Appl Phys Lett 2010, 96:013508.

18. Ye Y, Dai L, Wu PC, Liu C, Sun T, Ma RM,Qin GG: Schottky junction photovoltaic devices based on CdS single nanobelts. Nanotechonology 2009, 20:375202.

19. Jin YZ, Wang JP, Sun BQ, Blakesley JC, Greenham NC: Solution-processed ultraviolet photodetectors based on colloidal ZnO nanoparticles. Nano Lett 2008, 8:1649.

20. Yang WQ, Huo HB, Dai L, Ma RM, Liu SF, Ran GZ, Shen B, Lin CL,Qin GG: Electrical transport and electroluminescence properties of n-ZnO single nanowires. Nanotechnology 2006, 17:4868.

21. Ye Y, Dai Y, Dai L, Shi ZJ, Liu N, Wang F, Fu L, Peng RM, Wen XN, Chen ZJ, Liu ZF, Qin GG:High-performance single CdS nanowire (nanobelt) Schottky junction solar cells with Au/graphene Schottky electrodes. ACS Appl Mater Interfaces 2010, 2:3406.

22. Ye Y, Ma YG, Yue S, Dai L, Meng H, Li Z, Tong LM, Qin GG: Lasing of CdSe/SiO2 nanocables synthesized by the facile chemical vapor deposition method. Nanoscale 2011, 3:3072.

23. Gan L, Liu S, Li DN, Gu H, Cao Y, Shen Q, Wang ZX, Wang Q, Guo XF: Facile fabrication of the crossed nanotube-graphene junctions. Acta Phys – Chim Sin 2010, 26:1151.

24. Reina A, Jia XT, Ho J, Nezich D, Son H, Bulovic V, Dresselhaus MS, Kong J: Large area, few-layer graphene films on arbitrary substrates by chemical vapor deposition. Nano Lett 2010, 4:2689.

25. Castro Neto AH, Guinea F, Peres NMR, Novoselov KS, Geim AK: The electronic properties of graphene. Rev Mod Phys 2009, 81:109.

26. Thomas D, Boettcher J, Burghard M, Kern K: Photocurrent distribution in graphene-CdS nanowire devices. Small 1868, 2010:6.

27. Liu C, Dai L, Ye Y, Sun T, Peng RM, Wen XN, Wu PC, Qin GG: High-efficiency color tunable n-CdSxSe1-x/p+-Si parallel-nanobelts heterojunction light-emitting diodes. J Mater Chem 2010, 20:5011.

28. Ma RM, Wei XL, Dai L, Liu SF, Chen T, Yue S, Li Z, Chen Q, Qin GG: Light coupling and modulation in coupled nanowire ring-Fabry-Pérot cavity. Nano Lett 2009, 9:2679.

29. Flynn RA, Kim CS, Vurgaftman I, Kim M, Meyer JR, Mäkinen AJ, Bussmann K, Cheng L, Choa FS, Long JP: A room-temperature semiconductor spaser operating near 1.5 μm. Opt Express 2011, 19:8954.

30. Zhong ZH, Qian F, Wang DL, Lieber CM: Synthesis of p-type gallium nitride nanowires for electronic and photonic nanodevices. Nano Lett 2003, 3:343.

31. Gudiksen MS, Lauhon LJ, Wang JF, Smith DC, Lieber CM: Growth of nanowire superlattice structures for nanoscale photonics and electronics. Nature 2002, 415:617.

Chapter 4

GaInNAs-Based Hellish-Vertical Cavity Semiconductor Optical Amplifier for 1.3 μm Operation

Faten Adel Ismail Chaqmaqchee[1], Simone Mazzucato[1], Murat Oduncuoglu[1,2], Naci Balkan[1], Yun Sun[1], Mustafa Gunes[1], Maxime Hugues[3], and Mark Hopkinson[3]

[1]School of Computer Science and Electronic Engineering, University of Essex, Colchester CO4 3SQ, UK

[2]Department of Physics, Faculty of Science and Art, University of Kilis 7 Aralik, Kilis, Turkey

[3]Department of Electronic and Electrical Engineering, University of Sheffield, Sheffield S1 3JD, UK

ABSTRACT

Hot electron light emission and lasing in semiconductor heterostructure (Hellish) devices are surface emitters the operation of which is based on the longitudinal injection of electrons and holes in the active region. These devices can be designed to be used as vertical cavity surface emitting laser or, as in this study, as a vertical cavity semiconductor optical amplifier (VCSOA). This study investigates the prospects for a Hellish VCSOA based on GaInNAs/GaAs material for operation in the 1.3-μm wavelength range. Hellish VCSOAs have increased functionality, and use undoped distributed Bragg reflectors; and this coupled with direct injection into the active region is expected to yield improvements in the gain and bandwidth. The design of the Hellish VCSOA is based on the transfer matrix method and the optical field distribution within the structure, where the determination of the position of quantum wells is crucial. A full assessment of Hellish VCSOAs has been performed in a device with eleven layers of $Ga_{0.35}In_{0.65}N_{0.02}As_{0.08}$/GaAs quantum wells (QWs) in the active region. It was characterised through I-V, L-V and by spectral photoluminescence, electroluminescence and electro-photoluminescence as a function of temperature and applied bias. Cavity resonance and gain peak curves have been calculated at different temperatures. Good agreement between experimental and theoretical results has been obtained.

INTRODUCTION

III-V semiconductors are indispensable for today's optoelectronic devices, such as lasers modulators, photodetectors and optical amplifiers in optical fibre communication systems. One potentially important material for such applications is the quaternary alloy GaInNAs [1,2]. In the 1.3-μm optical communications window, GaInNAs may be grown pseudomorphically on GaAs, allowing the use of high quality AlAs/GaAs distributed Bragg reflectors (DBRs), with potential cost advantages compared to InP-based approaches. It can be used to fabricate several devices, among which vertical cavity semiconductor optical amplifiers (VCSOAs) are important components in optical fibre networks. They have improved performance over SOAs as they have inherent polarization insensitivity, lower noise figures, high-fibre coupling, easy chip testing and potential for integration into high-density two-dimensional arrays. Furthermore the narrower bandwidth of vertical cavity structures makes these devices good for filtering applications [3-6].

A VCSOA can be simply described as a vertical cavity surface emitting laser (VCSEL) operating in the linear regime below threshold, with a reduced number of top DBR layers. However, in this article, a novel VCSOA based on the Hellish structure as an alternative to conventional VCSOAs is investigated[7]. Hellish devices utilise the transport of non-equilibrium carriers parallel to the layers. Spontaneous emission of ultra bright Hellish structures has been demonstrated [8,9]. VCSEL operation was achieved by addiction of DBR layers [10-13]. That design is adapted in this study to make a GaInNAs-based Hellish-VCSOA structure, which differs from the conventional VCSEL by the reduced number of top DBR layers [14]. The structure is designed to operate in the 1.3-μm wavelength region via electrical pumping.

The authors demonstrate for the first time the operation of a Hellish VCSOA with a multiple quantum well (MQW) GaInNAs/GaAs active region, at temperatures between 77 and 300 K. Optical and electrical pumping (photoluminescence—PL, electroluminescence—EL) were used, and a 1.28-μm emission at room temperature was observed. By combining the two measurements, an electro-photoluminescence (EPL) technique was performed, from which light amplification is demonstrated. The authors also present the results of the reflectivity spectrum and cavity resonance calculations using the matrix formulation for multi-layer structures [15], and compare these with experimental results.

Experimental Results and Discussion

The structure of the investigated device, shown in Figure 1a, contains 11 layers of 6 nm-thick $Ga_{0.35}In_{0.65}N_{0.02}As_{0.08}$ quantum wells separated by 10 nm GaAs barriers. The use of MQWs, placed at the electric field antinode of $3\lambda/2$ cavity length, is done in order to provide optical gain (Figure1b). The active region is enclosed between two 150 nm-thick doped cladding layers Si-doped ($n = 1 \times 10^{17}$ cm^{-3}) on the bottom side, and C-doped ($p = 1 \times 10^{17}$ cm^{-3}) on the top side. The structure is sandwiched between two DBRs. The bottom DBR has 20.5 pairs of AlAs/GaAs quarter-wave stacks and provides a reflectivity in excess of 99% at 1.3-μm. On the other side, the top DBR has six pairs of AlAs/GaAs quarter-wave stacks giving around 60% reflectivity. This is lower than the bottom DBR, thus allowing light emission from the top surface. Both DBRs are undoped except for the first bottom AlAs/GaAs period which is 1×10^{17} cm^{-3} doped.

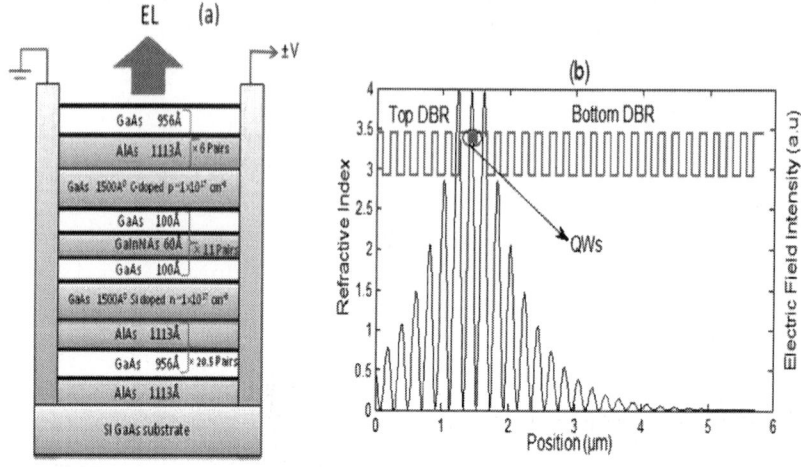

Figure 1: Schematic diagram illustrates (a) the layer structure of the simple bar Hellish-VCSOA and (b) the refractive index profile and distribution of the electric filed intensity across the sample, in which the QWs are situated at the antinode of the electric field, i.e. where it reaches its maximum intensity.

Ohmic contacts are formed by diffusing Au/GeAu/Ni/Au through all the layers and into the substrate, defining a simple bar-shaped sample, with 1-mm contact separation and 4.5-mm width. This is done by annealing the contacts for 60 s at 430°C.Once fabricated, the device is electrically biased with positive voltage pulses 390-ns duration and a 3-ms repetition rate. The duty cycle is small enough to prevent damage by excessive Joule heating. The applied electric field is varied from 0.01 to 1 kV/cm. Figure 2 shows the current-voltage (I-V) characteristics at 77 and 300 K. The sample exhibits ohmic behaviour at electric fields below 600 and 900 V/cm at 77 and 300 K, respectively. The small deviation from ohmic behaviour is an indication of carrier heating [16,17].

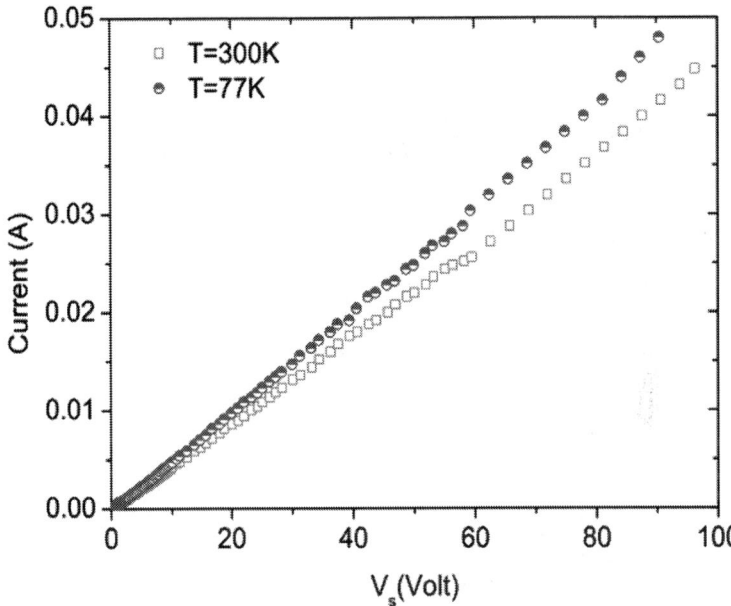

Figure 2: I-V characteristics of simple bar Hellish-VCSOA at 77 and 300 K .

The operation of Hellish device is based on the longitudinal injection of electron and hole pairs in their respective channels, due to the diffusion of both top contacts through all layers. Without the applied electric field, if the sample is illuminated, photogenerated carriers will eventually recombine radiatively in the QW without drifting laterally in the longitudinal channels. On the other hand, when the device is biased, the energy bands tilt up, with the degree of tilting being proportional to the applied voltage. At low bias, a quasi-flat region is established by the tilted energy bands, and a small number of carriers are then able to drift diagonally into the p-n junction. This is illustrated in Figure3. With an increase in the electric field, the energy bands will tilt up more, so that more carriers will flow into the active region, enhancing the emitted light. In view of the operational diagram depicted in Figure 3, the application of a negative bias results in a tilting and the diffusion of the holes to the region where electrons are injected, and recombination occurs in the vicinity of the cathode. This allows for spatial confinement and control of the light emission area. Luminescence from the opposite site (anode) appears by inverting the bias polarity [16].

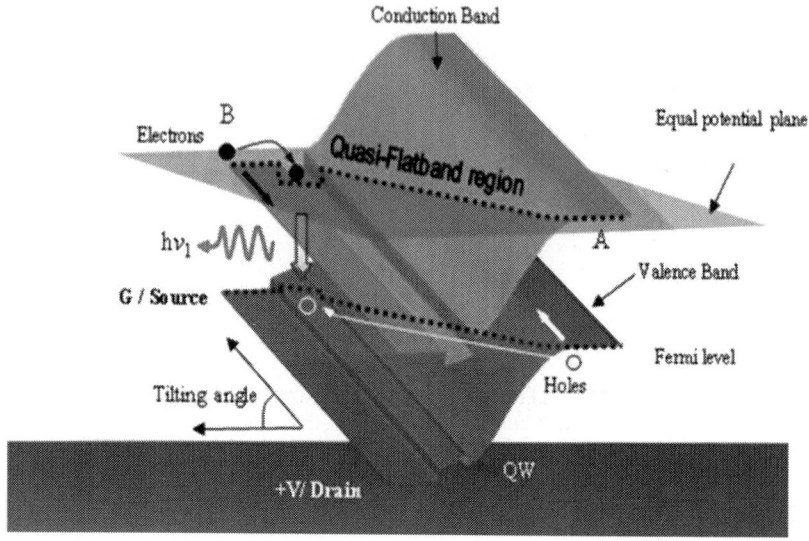

Figure 3: Schematic diagram to illustrate the emission of light under quasi-flat band region condition [16].

Experiments have been carried out using PL, EL and EPL techniques at different temperatures between liquid nitrogen and room temperature. The experimental arrangement for these techniques is illustrated in Figure 4.

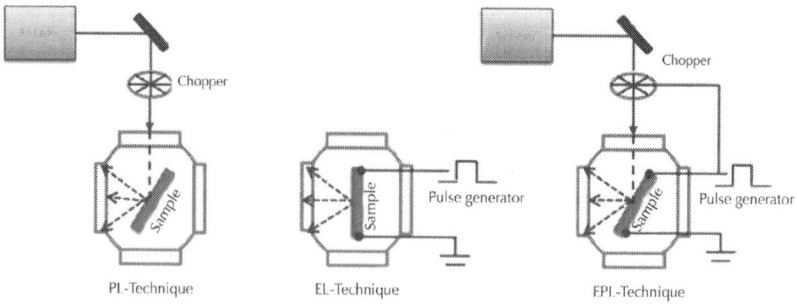

Figure 4: PL, EL and EPL experimental arrangement.

In PL and EPL, the optical excitation source is a CW Argon laser operating at 488-nm wavelength with 20-mW output power. The laser beam is chopped using a mechanical chopper and directed to the sample surface. The emitted light is dispersed by a Bentham M300 1/3 m monochromator and collected with a cooled InGaAs photomultiplier. The outcoming electrical signal is sent to a Gated Integrator & Boxcar Averager Module (Stanford Research Systems, model SR250) or a lock-in amplifier (Stanford Research Systems, model SR830) according to the experiment performed.

Figure 5 shows the integrated emission light from the device as a function of applied electric field. The threshold light emission varies between 110 and 290 V/cm according to the sample temperature. Above the threshold, the integrated EL increases linearly with applied electric field.

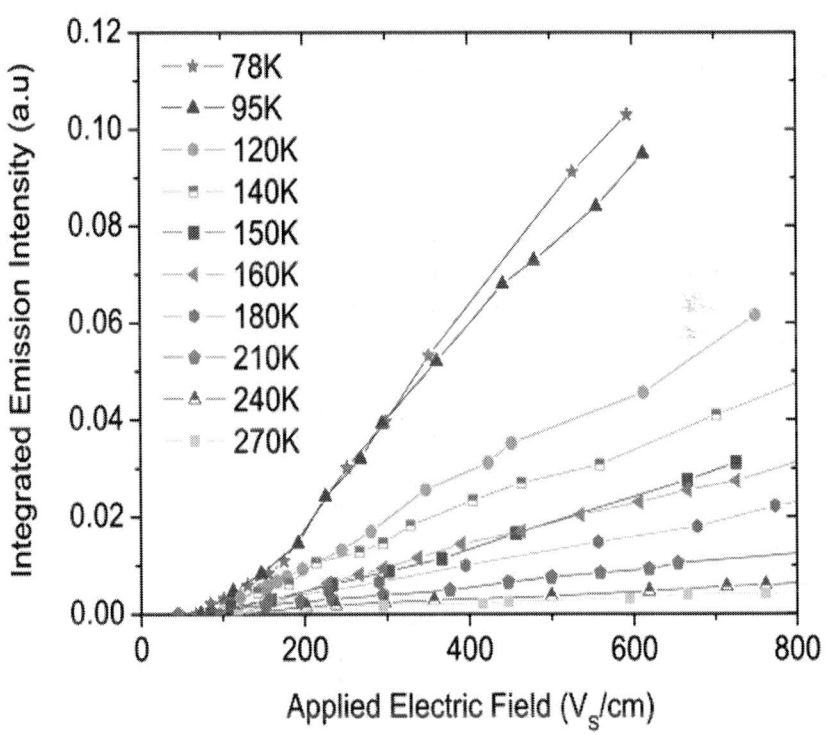

Figure 5: Integrated EL intensity versus applied electric field at various temperatures.

Figure 6 shows the PL spectra measured at different temperatures. The PL peak red-shifts from 1245 nm at 77 K to 1270 nm at 300 K.

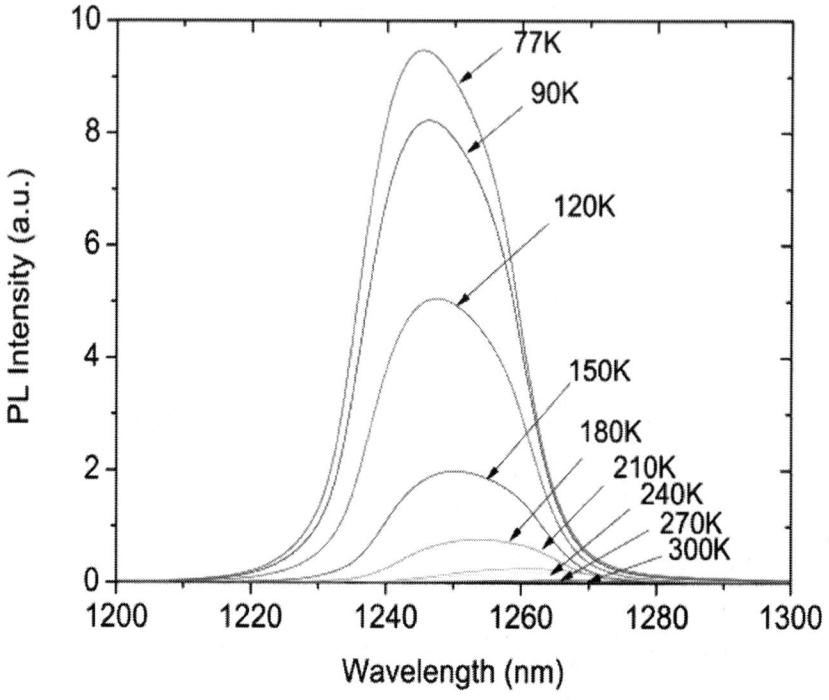

Figure 6: PL spectra measured at different temperatures.

Spectral EL is also measured with applied voltage pulses of amplitude between 0.3 and 100 V, where the pulse duration is kept at about 390 ns. The EL spectra are obtained at different temperatures between 80 and 300 K, and according to Figure 7, it shows a broad spectra. Approximately, the EL spectrum shifts in wavelength from 1239 nm at 80 K to 1281 nm at 300 K.

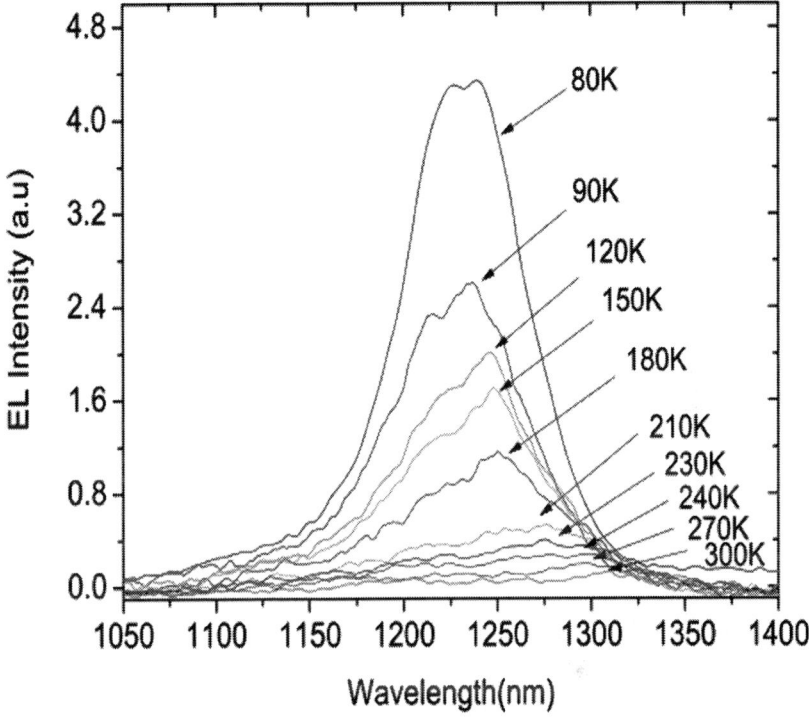

Figure 7: EL spectra measured at a fixed bias voltage of 97 V corresponding to an electric field of 0.97 kV/cm.

There is good agreement between the EL and PL peak positions. However, the EL emission is considerably broader than the PL. This observation is attributed to growth non-uniformities and material fluctuations. PL is measured from a small spot (excitation spot size 0.5 mm²), while the EL is collected from the whole sample surface. Therefore, the EL may be expected to be broader if the QWs and/or DBRs width have monolayer fluctuations. In order to prove this, the PL at different spots on the sample (Figure 8) was measured and the reflectivity spectrum for small fluctuations in the thickness of the layers in the cavity of around 2 nm (Figure 9) was calculated. The effect of layer fluctuations is clear.

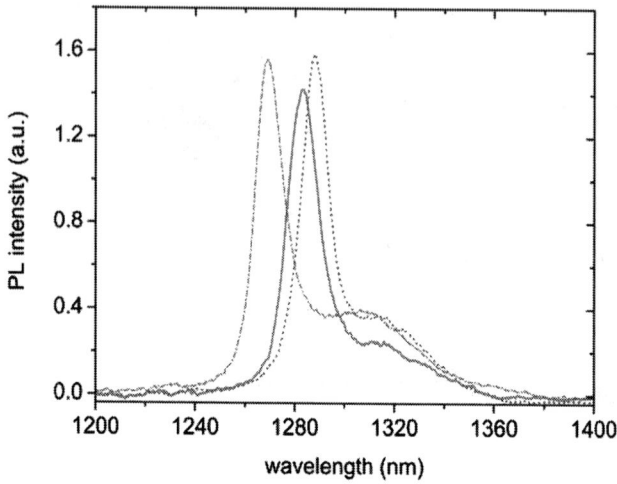

Figure 8: Room temperature PL spectra taken at different laser spot positions across the sample, showing an approximate 20-nm uncertainty in the peak position.

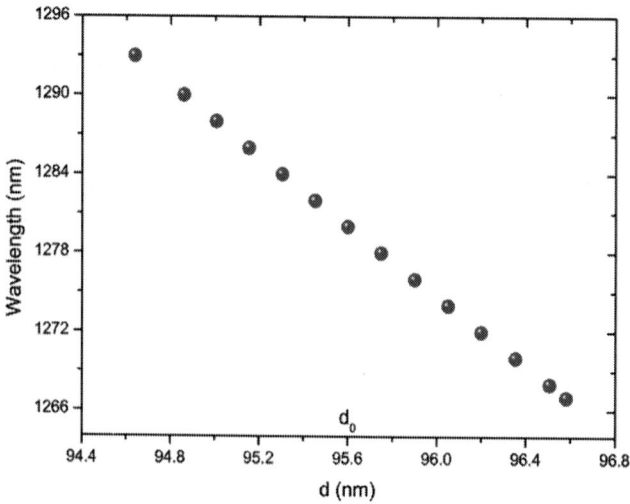

Figure 9: Dependence of the cavity resonance position with ±1 nm fluctuations of the GaAs thickness in the DBR (d_0 is the nominal GaAs DBR layer thickness) but weaker behaviour takes place when fluctuations occur in the AlAs layers.

The temperature dependence of EL and PL peaks and the cavity resonance are plotted in Figure 10, together with the active material bandgap energy curve [18]. The behaviour differs extremely from the change of the GaInNAs/GaAs bandgap energy with temperature. Theoretically, a red shift of the active material peak wavelength at a rate of 0.38 nm/K was predicted, while the resonance cavity moves with temperature at 0.18 nm/K. At these rates, the optimum operating temperature for this device will be at around 220 K, where the maximum peak material gain coincides with the DBR resonance cavity position.

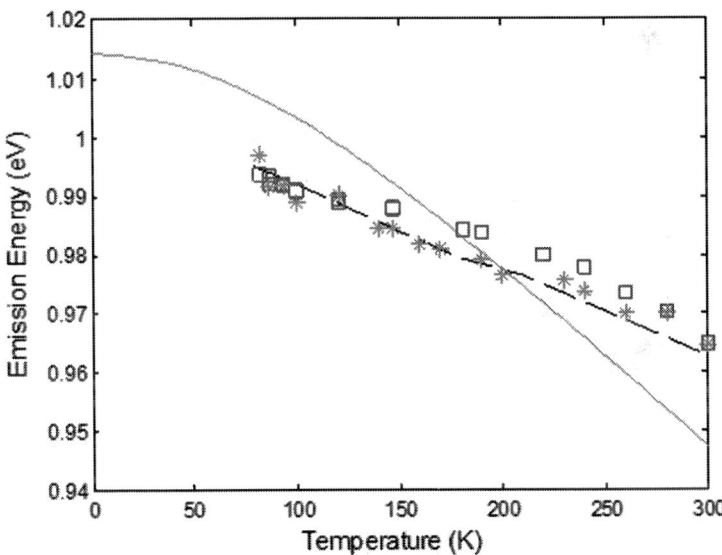

Figure 10: Continuous line represents the calculated temperature dependence of bandgap energy for the device active area (GaInNAs/GaAs QW) using the BAC model, while the expected cavity resonance position is plotted with a dashed line and finally the scattered points represent the experimental data for PL (asterisks) and EL (squares).

The EPL technique was performed by combining the two experimental techniques, namely PL and EL. In order to synchronise optical and electrical pulses, the pulse generator is triggered by a mechanical chopper. PL, EL and EPL spectra for Hellish-VCSOA are measured as a function of temperature. In both EL and EPL, the electric field was kept constant at 0.7 kV/cm.

In Figure 11, the T = 87 K EL, PL, EPL spectra and the sum of EL and PL are plotted. The PL spectrum presents a broad peak at around 1250-nm wavelength and a full-width-at-half-maximum of 13 meV. As stated before, it corresponds to the overlap of the active region gain spectrum and the cavity resonance reflectivity that filters and narrows the emission. Variations in the peak position are ascribable to fluctuations in the cavity resonance. The EL spectrum measured at the same temperature shows the emission peak at around 1.03 eV and by comparing the SUM (EL + PL) and EPL spectrum, the presence of optical gain was clearly visible. Signal amplification occurs when both electrical and optical inputs are applied.

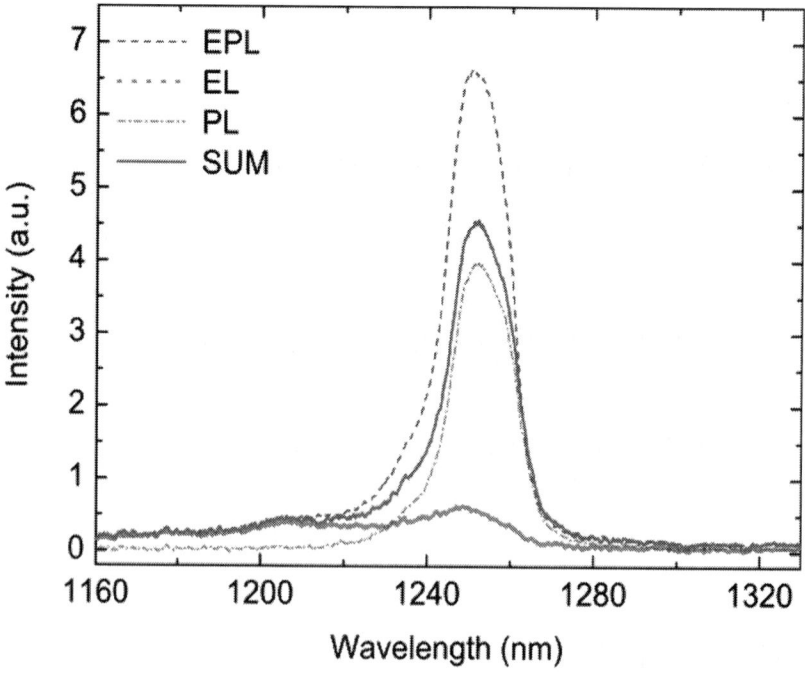

Figure 11: EL, PL, EPL and SUM (EL + PL) spectra at T = 87 K.

This investigation was focussed on the gain at room temperature. The integrated intensities of PL, EL and EPL, together with the calculated SUM (EL + PL) and gain are plotted in Figure 12, as function of applied voltage, up to 800 V/cm, with laser excitation power of 10 mW. Finally, Figure 13 displays the evolution of the gain with applied voltage, which reaches its maximum at around 50 V. It should be noted that the wavelength of the laser (λ = 488 nm) is very different from the cavity resonance position shown in Figure 10. Therefore, most of the excitation is lost through absorption. In order to give a quantitative value to the VCSOA gain, the PL gain is defined as the ratio of the PL peak when the device is electrically pumped to that when the device is not biased. This gain should not be confused with conventional VCSOA gain as ratio of output power to input power.

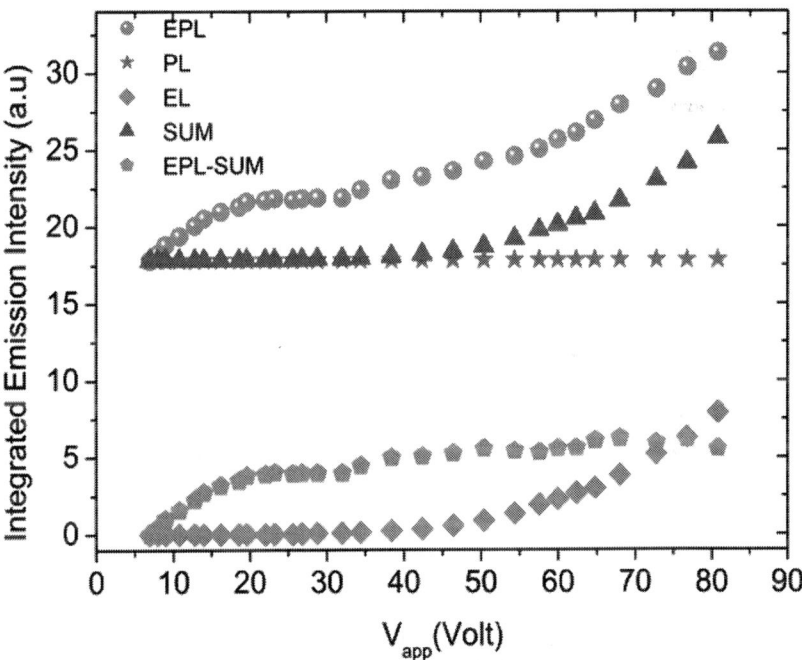

Figure 12: Integrated EPL, PL, EL, SUM (EL + PL) and EPL-SUM (EL + PL) measured as a function of applied voltages at T = 300 K.

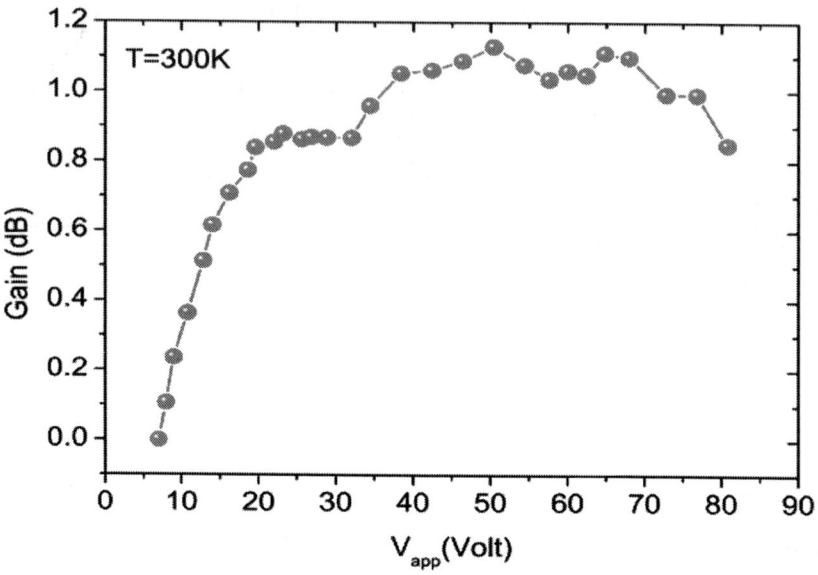

Figure 13: Gain characteristics are measured as a function of applied voltages at T = 300 K.

Further improvements in gain characteristic and device performance will be expected by optimising the Hellish-VCSOA structure for 1.3-μm application via electrically pumping, and by reducing the device length so that the operating voltage will be much lower than the one used here.

CONCLUSIONS

Optical gain at $\lambda \sim 1.28$ μm is demonstrated in a Hellish-VCSOA device consisting of $Ga_{0.35}In_{0.65}N_{0.02}As_{0.08}$/GaAs QWs and AlAs/GaAs DBRs. The advantage of using such device is that longitudinal electric fields are applied parallel to active layer so that the current flows along the p andn layers without passing through the DBRs. The operation of the device is independent of the polarity of the applied electric field. The emission and amplification characteristics are investigated as a function of temperature and applied voltage. Thus, the Hellish-VCSOA is a good candidate for electrically pumped optical amplifier operating at around 1.3 μm.

AUTHORS' CONTRIBUTIONS

FAI, YS and MO designed the structure. MHu and MHo grew the sample according to the specifications. FAI fabricated the devices, carried out the experiments and the theoretical calculations, in collaboration with MO, YS, SM, and MG. FAI and SM wrote up the article. NB, is the inventor of the original device and the overall supervisor of the project.

All authors read and approved the final manuscript.

ACKNOWLEDGEMENTS

F.Al. Chaqmaqchee is grateful to the Ministry of Higher Education and Scientific Research of IRAQ for their financial support. M. Oduncuoglu is grateful to Kilis 7 Aralik University/Turkey research fund and the financial support provided by TUBITAK. The authors also acknowledge A. Boland-Thoms for technical assistance. Finally we are grateful to the COST Action MP0805 for providing the scientific platform for collaborative research.

REFERENCES

1. Kondow M, Uomi K, Niwa A, Kitatani T, Watahiki S, Yazawa Y: GaInNAs: A novel material for long-wavelength-range laser diodes with excellent high-temperature performance. Jpn J Appl Phys 1996, 35:1273.

2. Buyanova IA, Chen WM, Monemar B: Electronic properties of Ga(In)NAs alloys. MRS Internet J Nitride Semicond Res 2001, 6:2.

3. Piprek J, Bjorlin ES, Bowers JE: Optical gain-bandwidth product of vertical-cavity laser amplifiers. Electron Lett 2001, 37:298.

4. Karim A, Bjorlin S, Piprek J, Bowers JE: Long-wavelength vertical-cavity lasers and amplifiers. IEEE J Sel Top Quantum Electron 2000, 6:1244.

5. Bjorlin S, Riou B, Keating A, Abraham P, Chiu Yi-J, Piprek J, Bowers JE: 1.3-μm vertical-cavity amplifier. IEEE Photonics Technol Lett 2000, 12:951.

6.   Bjorlin S, Abraham P, Pasquariello D, Piprek J, Chiu Yi-J, Bowers JE: High gain, high efficiency vertical-cavity semiconductor optical amplifiers. Indium Phosphide and Related Materials Conference 2002., 307

7.   Balkan N, Serpenguzel A, O'Brien-Davies A, Sokmen I, Hepburn C, Potter R, Adams MJ, Roberts JS: VCSEL structure hot electron light emitter. Mater Sci Eng 2000, B74:96.

8.   Straw A, Balkan N, O'Brien A, da Cunha A, Gupta R: Hot electron light-emitting and lasing semiconductor heterostructures--type 1. Superlatt Microstruct 1995, 18:33.

9.   O'Brien A, Balkan N, Boland-Thoms A, Adams M, Bek A, Serpenguzel A, Aydinli A, Roberts J:Super-radiant surface emission from a quasi-cavity hot electron light emitter. Opt Quantum Electron 1999, 31:183.

10.   Balkan N, Sokmen I, O'Brien A, Potter R, Hepburn C, Boland-Thoms A, Adams MJ, Roberts J:Hot electron VCSEL. Proc SPIE 1999, 3625:336.

11.   Balkan N, O'Brien A, Boland-Thoms A, Potter R, Poolton N, Adams M, Masum J, Bek A, Serpenguzel A, Aydinli A, Roberts J: The operation of a novel hot electron Vertical Cavity Surface Emitting Laser. Proc SPIE 1998, 3228:162.

12.   Erol A, Balkan N, Arıkan MC, Serpenguzel A, Roberts J: Temperature Dependence of the Threshold Electric Field in Hot Electron VCSELs. IEE Proc Optoelectron 2003, 150:535.

13.   Sceats R, Balkan N: Hot Electron Light Emission at 1.3 µm from a GaInAsP/InP structure with distributed Bragg reflectors. Phys Status Solidi 2003, 198:495.

14.   Wah JY, Loubet N, Mazzucato S, Balkan N: Bi-directional field effect light emitting and absorbing heterojunction with $Ga_{0.8}In_{0.2}N_{0.015}As_{0.985}$ at 1250 nm. IEE Proc Optoelectron 2003, 150:72.

15.   Yeh P: Chapt. 5 of Optical Waves and Layered Media. New York: John Wiley & Sons; 1991.

16.   Wah JY, Balkan N: Low field operation of hot electron light emitting devices: quasi-flat-band model. IEE Proc Optoelectron 2004, 151:482.

17. Wah JY, Balkan N, Potter RJ, Roberts JS: The operation of a Wavelength Converter based on Field Effect Light emitter and absorber heterojunction. Phys Status Solidi A 2003, 196:503.

18. Potter RJ, Balkan N: Optical Properties of GaInNAs and GaNAs QWs. J Phys Condens Matter 2004, 16:3387.

Chapter 5

Application of Metal-Semiconductor-Metal Photodetector in High-Speed Optical Communication Systems

Farzaneh Fadakar Masouleh[1], and Narottam Das[2, 3]

[1]Physics Department, University of Guilan, Rasht, Guilan, Iran

[2]Department of Electrical and Computer Engineering, Curtin University, Perth, WA, Australia

[3]Department of Electrical and Computer Engineering, Curtin University, Miri, Sarawak, Malaysia

INTRODUCTION

Recently, there is a very strong interest towards the miniaturization of optical and electrical components with faster and more efficient performance which has incorporated new capabilities in various aspects including high-speed telecommunication systems [1]. Optical communication technology has greatly developed during recent years that affect all areas of the modern telecommunication systems. Photodetectors along with optical sources and fibers are regarded as integral part of all optical fiber communication systems [2]. The high bandwidth and/or the gain of photodetectors with the wavelength in the near-infrared region (0.8 to 1.6 µm) are quite important because of their ideal commercial and industrial usage in optical fiber communication systems. The photodetectors are known as optoelectronic devices that can convert the absorbed optical energy into electrical energy which usually appears as a photocurrent that can be used by telephone systems, computers, or other terminals at transmitting and receiving part of the communication systems [3]. There are many types of photodetectors depending on their application in optical communication systems, imaging systems, and so on. The sensitivity of detectors varies through different optical spectra, such as the infrared and ultraviolet. Semiconductor detectors are commonly used in optical fiber communication systems because they rely on internal photoelectric effect but there is no photoemission effect. They either work in photovoltaic mode as solar cells or in photoconductive mode as revers biased photodetectors [4]. Metal-semiconductor-metal photodetectors (MSM-PDs), positive-intrinsic-negative (pin) photodetectors, avalanche photodiodes and heterojunction PDs as solid state semiconductor devices are most widely used in high-speed optical communication systems. The MSM-PDs are attractive devices compared with other photodetector structures, for their remarkable high sensitivity-bandwidth product, ease of fabrication and ease of integration with other components into monolithic receiver circuits. The MSM-PD consists of two identical Schottky contacts with interdigitated electrode configuration on top of an undoped semiconductor substrate, one of the contacts being forward and the other reverse biased. With fabrication of interdigitated electrodes the closely spaced fingers provide smaller transit time for the carriers as well as allowing a larger photo-absorption area for the device [5, 6]. The two Schottky barriers associated with the presence of contacts block current flow from the metal to the semiconductor which is the cause for the extremely low dark current observed in MSM-PDs. One important feature of the MSM photodetector is its low capacitance compared with a pin photodetector, (with an intrinsic region (i.e., undoped semiconductor) in between the n-and p-type semiconductors). The capacitance of a MSM photodetector with interdigitated electrodes is always smaller than that of a pin photodetector of equal light sensitive area and leads to very high-performance operation.

High speed photodetection manifests an exciting new paradigm for modern telecommunication systems. Advanced or modern optical systems, i.e., the ultrahigh-speed optical telecommunication systems, such as any typical fiber optic communication system consist of a transmitter, a data transmission media or channel (usually optical fiber, waveguide, and free space air-gap), and a receiver (pin photodiodes and photodetectors). The major part of the optical transmitter is a light source (laser or light-emitting diode (LED)), whose function is to convert an information signal from its electrical form into light. The photodetectors as an important part of receiver are used to convert an optical information signal back into an electrical signal. For higher speed and broader bandwidth applications in optical communications and interconnects, high-performance optical receivers are required and the MSM photodetector as the heart of optical receiver has many advantages such as wide bandwidth and low capacitance. MSM-PDs are promising candidates in optoelectronic integrated circuits. Also they have fast time response and very low dark current, as compared with other types of photodetectors.

The new field of plasmonics has received particular attention due to unique optical features of nano scale architectures in noble metals particularly silver and gold. The coherent oscillations of electrons as surface plasmons are strongly localized in the nanoscale at the metal-dielectric interface, and metal nanoparticles. When the losses are small enough, the surface plasmon resonances can occur. The metals with good quality for plasmonic applications should satisfy two properties such as $\varepsilon\prime m < -1$, and $\varepsilon\prime\prime m << |\varepsilon\prime m|$, where $\varepsilon\prime m$, and $\varepsilon\prime\prime m$ are real and imaginary parts of metal dielectric permittivity, respectively [7].

Dielectric and magnetic properties of the noble metals with nano-textured structure can easily be determined by implementation of Lorentz-Drude model [8]. Below plasma frequency, the good conductors like silver (Ag) and gold (Au) have negative values for the real part of complex dielectric constant. Therefore, it is necessary to define an appropriate model to specify the dielectric properties of the materials. A complex dielectric function for some metals and surface plasmas which have good agreement with the experimentally measured results can be expressed in the following form [9]:

$$\varepsilon_r(\omega) = \varepsilon_r^f(\omega) + \varepsilon_r^b(\omega)$$

(1)

where, the term $\varepsilon_r(\omega)$ is the complex dielectric function for metals, $\varepsilon_r^f(\omega)$ is referred to as free electron effects, and $\varepsilon_r^b(\omega)$ is associated with bound electron effects. This model takes both the intraband, $\varepsilon_r^f(\omega)$, and interband, $\varepsilon_r^b(\omega)$, effects into the account for simulations. The former, Drude model, can describe the transport properties of electrons in good conductors and the later, Lorentz model, is a semi-quantum model describing bound electron effects. The Drude and Lorentz model in frequency domain can be written in the following form of (2) and (3), respectively.

$$\varepsilon_r^f(\omega) = 1 + \frac{\Omega_p^2}{i\omega\Gamma_0 - \omega^2}$$

(2)

$$\varepsilon_r^b(\omega) = \sum_{m=1}^{M} \frac{G_m \omega_p^2}{\omega_m^2 - \omega^2 + i\omega\Gamma_m}$$

(3)

where, $\Omega_p = G_0^{1/2}\omega_p$ is the plasma frequency associated with intraband transitions with oscillator strength G_0 and damping constant Γ_0, while m is the number of oscillators with frequency ω_m and $1/\Gamma_m$ is the oscillator lifetime [10]. The following equation accounts for the complex index of refraction and dielectric constant of materials and can be represented as a combined form [11]:

$$\varepsilon_r(\omega) = \varepsilon_{r,\infty} + \sum_{m=0}^{M} \frac{G_m \Omega_m^2}{\omega_m^2 - \omega^2 + i\omega\Gamma_m}$$

(4)

where, $\varepsilon_{r,\infty}$ is the relative permittivity at infinite frequency, Gm is the strength of each resonance term, Ω_m is the plasma frequency, ω_m is the resonant frequency, and Γ_m is the damping factor or the collision frequency.

The frequency dependent Gold dielectric permittivity is complex and is obtained from Lorentz-Drude model, $\varepsilon_m = \varepsilon'_m + i\varepsilon''_m$. It consists of a large negative real part $\varepsilon'm$ and a small positive imaginary part $\varepsilon''m$ responsible for light absorption. The letter i is an imaginary unity. Gold dielectric constant varies for different frequency ranges.

For our simulation of plasmonic-base MSM-PD, the Lorentz-Drude model for gold is solved with 6 resonant frequencies (multi-pole dispersion). The simulation is performed over a constant plasma frequency which depends on the density of charge carriers with the amount of 0.137188E+17. The resonant frequency changes according to the resonance strength. The Lorentz-Drude material parameters are listed in the following Table-I.

Table 1: Lorentz-Drude parameters for gold (measured in radians per second)

Term	Strength	Resonant Frequency	Damping Frequency
0	0.7600	0.000000E + 00	0.805202E + 14
1	0.0240	0.630488E + 15	0.366139E + 15
2	0.0100	0.126098E + 16	0.524141E + 15
3	0.0710	0.451065E + 16	0.132175E + 16
4	0.6010	0.653885E + 16	0.378901E + 16
5	4.3840	0.202364E + 17	0.336362E + 16

Investigation on the optical properties of the nano-structures is crucial for the design of optical devices. Next generation information and communication technologies are under direct influence of nano-plasmonic devices because of their unique properties that hold promise for potential applications in various fields of technology, properties such as overcoming the diffraction limit, efficiency in confinement of light at subwavelength scale, and ultrahigh speed signal transport with the same order as the speed of light make them vigorous devices.

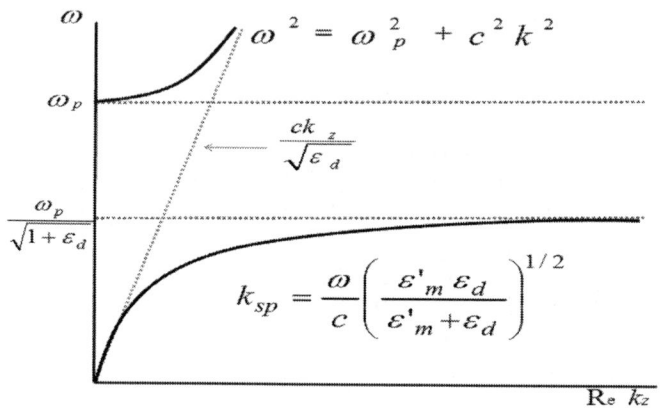

Figure 1: The dispersion relation of non-radiative SPs on the right of the light line and dispersion relation of radiative SPs on the left of the light line.

There has been a considerable and growing interest to clarify all phenomena corresponding to the improvement in light absorption via nano-scale structuring recently [12-15]. Over the decade, an interesting effect of light interacting with metallic structures has been revealed. For sufficiently small nano-grating period, higher order diffractions are suppressed and only the zero order diffraction is present. The main motivation is to come up with a practical method to avoid the undesirable light reflection from the surface of the electrodes and enhance the light transmission efficiency through the subwavelength aperture.

It is useful to describe the SP dispersion relation as shown in Fig. 1. To have propagating bound SPs, the wave vector component must be real along the interface. Any light cannot be used for SP generation, because the real part of the wave vector in the z-direction exceeds that of free light ($\omega = ck$) as shown in Fig. 1 and a coupling mechanism is required. In the presence of a transverse magnetic (TM) polarized light, the SPs exist along a metal-dielectric interface. There are several techniques to excite the SP waves, such as prism coupling and grating coupling. For the case of semiconductors, prism coupling is not very much advantageous. The attention should be paid to prism refractive index which is hardly higher than popular semiconductors used for corresponding researches [16, 17].

NANO-GRATINGS STRUCTURES AND OPERATION PRINCIPLES

The metallic nano-gratings can exhibit absorption anomalies. One of these particularly remarkable anomalies is the surface plasmon polariton (SPP) excitations and is observed for p-polarized light only. The incident light illuminated on top of the one dimensional metallic nano-gratings and subwavelength slits is converted into the propagating SPPs that can absorb the light efficiently in extremely thin (10-nm to more than 100-nm's thick) layers. The coupling of light to the structure in the form of the SPPs is obtained with the utility of periodicity. There are two mechanisms to produce the transmission of light in one-dimensional (1-D) metal nano-gratings with narrow slits which are the excitation of horizontal and vertical surface resonances however there is not always a clear distinction between these two ways of transferring light from the upper surface to the lower one and that their existence strongly depends on the gratings geometry. The horizontal surface resonances are cavity modes excited by the periodic structure of the nano-gratings at the grating's interfaces. Two vertical resonances existing on the slit walls can compose a fundamental waveguide mode which satisfies the Fabry-Perot condition and reflects repeatedly from both ends of the slit. Therefore, the vertical surface resonances have relations with Fabry–Perot resonances of the fundamental TM guided wave in the slits [18], and the guided modes can be excited along vertical direction, evanescent coupling mechanism accounts for this activation [19]. The responsibility of these two effects on extraordinary optical transmission (EOT) is still under debate. However, it is indicated that both mechanisms are important in the EOT [20].

Nowadays, it is clear that the real metals play a very important role in EOT, because their effect is essential for the SP excitation. The condition for the existence of the SP on a flat air-metal interface, $\varepsilon'_m < -1$, $\varepsilon''_m << |\varepsilon'_m|$, is fulfilled for many metals, including the silver and gold.

Concentrating the light in small areas with the assistance of extremely thin layers (plasmonic lenses), the EOT is a unique phenomenon which is introduced as the SP assisted multiple diffractions, coupling the SPs into the aperture, and conversion back into near-field at the exit side of the aperture. The role of SP modes is essential to clarify the EOT of light through the slits. In the EOT, the aperture transmits more light than the standard aperture theory. The EOT can occur whenever some specific conditions are satisfied, i.e., the slit width must be much smaller than the incident light wavelength, the periodicity has to be in the range of wavelength and it can lead to outstanding results if the light is normally incident to the structure's surface. Subwavelength apertures have also been used to concentrate light efficiently into the deep subwavelength regions.

A considerable development has been reported for the EOT through metallic gratings with very narrow slits. The SPPs and subwavelength slits in metallic thin films are always involved with this phenomenon but the details are still under investigation [21, 23]. The SPs can be efficiently excited in the nano-structured noble metals since they almost have free-electron behavior. The noble metal nano-textured structures have special properties to produce localized regions of high energy concentration and show larger enhancement for EOT. This effect and its underlying mechanism have important applications in photolithography and near-field microscopy.

Specifically, in this chapter we present the effects of nano-grating structures on the MSM photodetector performance. Optimization of subwavelength nano-gratings shape and pitch is very much important to generate the zero-order diffraction waves. Thus, the subwavelength nano-gratings can be represented as a homogeneous medium with optical properties determined by the nano-grating geometry. When the nano-grating period is within the order of the light wavelength, the light wave may be resonant and reflects into the structure, hence the higher diffraction orders will be suppressed and resonant reflection occurs for the zero-order diffraction waves.

MODELING AND CHARACTERIZATION OF PLASMONIC-BASED MSM-PDS

For proper design of MSM photodetector structure, it is necessary to define an appropriate model to specify the dielectric properties of the materials. In our modeling, semiconductor substrate is made up of gallium arsenide (GaAs) which is a direct band gap compound semiconductor. Usually, the GaAs substrate is preferred for the design of electronic and photonic devices, because it has unique electrical properties. Being a direct bandgap semiconductor, it can collect and emit light more efficiently than indirect band gap semiconductors such as Si and Ge. The GaAs substrate as an isotropic material has a constant refractive index of $3.666+i6.12\times10^{-2}$ in our simulation. GaAs is suitable for infrared range applications and optical fiber communication because of its wide bandgap and fast electron conduction. It also has short absorption length that enables the detector to combine wide range bandwidth with good responsivity characteristics. The GaAs has a conduction-band structure that leads to fast electron conduction.

Periodic nano-structures can produce an efficient light transmission and absorption by excitation of surface plasmons (SPs). The continuing progress in plasmonic interaction with nano-structures and their outstanding effects in development of MSM-PDs design have developed a unique context for future-generation optoelectronic systems, such as, optical fiber communication systems.

The conventional MSM-PD is a symmetrical device equivalent to two back-to-back connected Schottky diodes on a semiconductor substrate. To create a Schottky junction, along with the shape and size of the interface, some essential properties, such as type and quality of metal and semiconductor must be satisfied.

A conventional MSM-PD structure is shown in Fig. 2. By impinging the light from the top on the conventional device surface, a considerable amount of illumination on top is reflected, so the light absorption inside the MSM photodetector (or substrate) is significantly reduced in device's active region. Fabrication of nano-gratings on the metal fingers of the conventional MSM-PD structure avoids this unwanted phenomenon (or light reflection), compare Fig. 2 and 3. Such nanostructure then can capture most of the reflecting light inside the active region of the device (or substrate). Hence, the individual subwavelength apertures can exhibit notable light transmission when surrounded by periodic nano-structures that harvest the externally incident light into the slit. This proposed mechanism is shown in Fig. 3.

By tailoring the electrodes structure surface with metallic nano-gratings, MSM-PDs can be modified for the light absorption and the modification process strongly depends on the corrugation parameters. Recently, different shapes of 1-D nano-structured surfaces have been developed in which noble metals, such as gold (Au) are used for nano-grating structuring. When an electric field (or a voltage) is applied between the electrodes and the device's active region is under illumination then the electric carriers (i.e., electrons and holes) are generated and drifted towards the opposite electrodes due to the electric field and can form a photocurrent. Improvement of recombination between electrons and holes can lead to enhance the light absorption in the subwavelength aperture region.

The TM polarized wave is required to excite the propagating SP waves, when the z-component of the light k-vector matches with the SP k-vector, because only the TM polarization has electric field component in the same direction as surface normal. The polarization is described with regard to the electric field configuration of the incident light with respect to the surface normal (angle β) as shown inFig. 3. The wave is purely TE polarized, when $\beta=\pi/2$, and purely TM polarized, when $\beta=0$. The angle ϕrepresents the direction of the plane of incidence, such that the nano-grating is in the classical mount in this model, where all the diffracted orders remain in the x-z plane and the plane of incidence is normal to the nano-grating grooves ($\phi=0$). Besides, when the grooves are parallel to the plane of incidence, the nano-grating is in the conical mount ($\phi=90$). In our simulation, we used the nano-grating period of =810-nm. The grating period has to be optimized to effectively couple surface plasmon resonances to the light wave and trigger the quality absorption process.

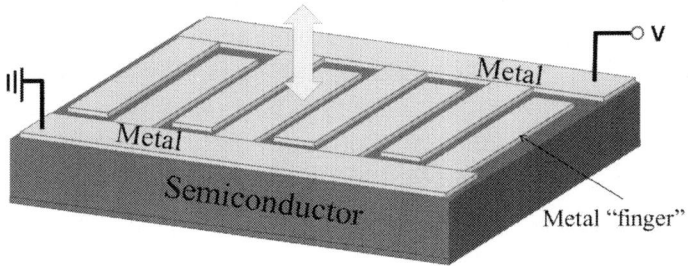

Figure 2: Schematic diagram of conventional MSM-PD structure with inter-digitated electrodes (metal fingers) and semiconductor substrate (GaAs). For this symmetric device under illumination, there are undesirable light reflections.

It is very important to build a new type of plasmonic-based MSM-PD with the characterization of high speed and high responsivity application which is useful in optical interconnect and communication systems. An MSM-PD speed in operation is intrinsically limited by the carrier transit time between the fingers or by the carrier recombination time. Although reducing the interdigitated electrodes spacing as well as scaling down the entire MSM-PD dimensions are common ways to increase the photodetector speed, as transit time is quite small, undesirable light reflection from the surface of the electrodes and the metal fingers shadowing the detection area which leads to decreased active area, lowers the MSM-PD responsivity. Employing metal nano-gratings on interdigitated electrodes demonstrates substantial transmission enhancement through the excitation and guidance of surface plasmon polaritons into the photodetector's semiconductor region [24].

PLASMONIC-BASED MSM-PD SIMULATION MODEL

Metallic nano-grating structures produce efficient and ultrafast photodetection properties. In this section, a plasmonic-based MSM-PD is introduced which utilizes electromagnetic and optical properties of nanostructures to enhance the light harvesting inside the device active region. The plasmonic-based MSM-PD consists of a semiconductor absorbing layer on which the interdigitated electrodes have been deposited to form two back-to-back connected Schottky diodes. The interdigitated electrodes are designed to resolve the conventional MSM-PD's degraded efficiency problem because of the metallic electrodes opacity. Also, the MSM-PDs are patterned by nano-gratings to improve the light capturing capacity into the device active region. Electromagnetic fields coupled to a charge density wave propagating at the metal-dielectric interface produce transverse-magnetic optical surface waves namely surface plasmon polaritons. The coupling condition for the SPPs is provided by metallic nano-gratings with optimized dimensions and geometry. This structure has the feasibility to create a plasmonic lens which utilizes plasmonic effects to produce surface plasmon resonances and funnel the energy toward the central focal point.

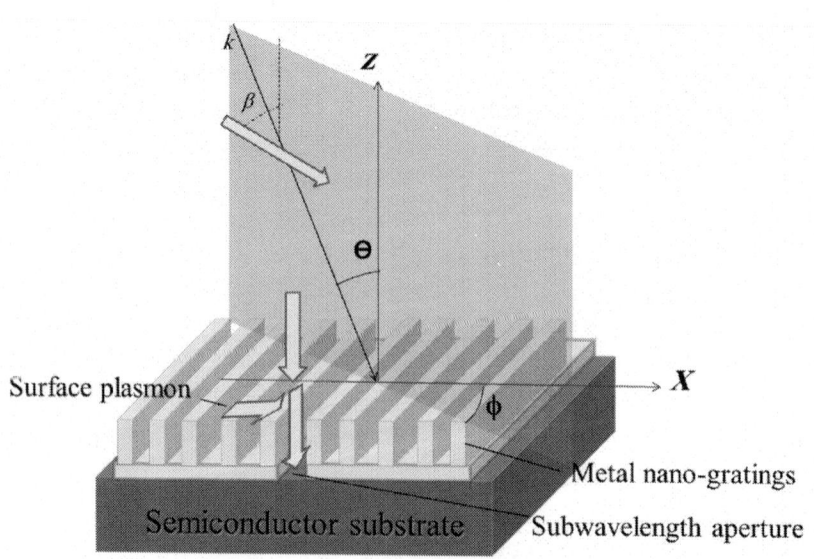

Figure 3: Schematic diagram of MSM-PD structure with rectangular shaped nano-gratings on top of the subwavelength slit. The subwavelength slit layer is just on top of the semiconductor (GaAs) substrates. The excited SPPs travel along the interface to reach the subwavelength slit/aperture.

The design of high performance plasmonic-based MSM-PD is shown in Fig. 3. A plasmonic-based MSM-PD structure has three separate regions or parts, namely, the incident region (the metal nano-gratings), the under layer and the subwavelength slit region, and the transmission region (the semiconductor substrate). The nano-scale gaps between the interdigitated electrodes in the MSM-PDs result in a huge increase in bandwidth and reduction in dark current, whereas the conventional pin photodetectors with similar sized active areas are unable to achieve that amount of light absorption.

In addition, the near-field characteristics and associated field enhancements can be achieved for periodic nano-structures and subwavelength apertures with aid of the SPPs interactions. This suggests that, properly designed metallic nano-grating grooves trigger surface plasmon polaritons under illumination and carry them toward the central slit. By using a subwavelength central slit, a well-directed source of light could be generated, an exciting development that is being pursued as a source for a variety of optical technologies. The light continuously re-emits from a very small area surrounding the central aperture which is associated with properties of the Fabry-Perot cavity resonances for symmetric SPP modes of the slit. Particularly, The SPPs which are supported by the active region of the device show a great potential of subwavelength photonic phenomenon. The excitation of the SPP waves causes a resonance absorption, which can be observed as partial or total absorption of the incident light. The absorption enhancement caused by the excitation of SPPs is associated with the incident photons and their interaction with the nano-gratings, while Fabry-Perot like resonances are included in transmission absorption process through the subwavelength slits.

In periodic subwavelength structures, the nano-gratings are deposited on top of the under layer from the same metal (such as gold). The absorption enhancement can be achieved by the SPP resonant excitations in the subwavelength region. Here, the FDTD method is used to specify the TM polarized plane wave via Poynting vector. The SPPs are evanescent waves, which are generated by the interaction between the surface electron densities and the electromagnetic fields whilst trapping the light power inside the surface. Since the SP modes have longer wave vectors than the light waves with the same energy, the SP waves are non-radiative on smooth metallic surfaces and cannot propagate in non-metallic media. One way to excite the SPPs is the nano-grating coupling technique in which the incident radiation is coupled into the SPPs using periodic surface corrugations with proper dimensions. The nano-grating grooves are perpendicular to the x-direction and its dimensions and geometry are optimized to couple the light near the design wavelength, that is, providing the missing momentum in order to make the SPPs propagate along the z-direction. In a metal-dielectric interface, the SPP wave vector matching condition for a metal nano-grating can be defined with some changes to the well-known prism resonance condition. Hence, in a metal-dielectric interface, the SPP propagating constant or wave vector matching condition for a metal nano-grating with the period of Λ is given by [22, 24]:

$$ k_{spp} = \frac{\omega}{c}\sin(\theta) \pm \frac{2\pi l}{\Lambda} = \frac{\omega}{c}\sqrt{\frac{\varepsilon'_m \varepsilon_d}{\varepsilon'_m + \varepsilon_d}} $$

(5)

where, ω is the angular frequency of the incident light wave, c is the speed of light in vacuum. l is an integer number i.e., l=1, 2, 3, ..., N and θ is the light angle of incidence to the device normal. This relation illustrates that the wave vector of a given frequency is smaller than the SPP wave vector, therefore the light wave vector should increase with the support of a coupling mechanism to provide SPPs which in this case is satisfied by nano-grating structures.

The wave vector has to be complex as the metal permittivity is complex, $\varepsilon_m = \varepsilon'_m + i\varepsilon''_m$. To trigger SPPs, the dielectric permittivity has to change sign in the metal-dielectric interface. The values of dielectric constant vary for different frequency ranges. To represent the influence of electric field in organizing electric charges and dipoles in the medium, we introduce electric displacement field, $D = \varepsilon_0 \varepsilon_r E$, between two isotropic media where, ε_0 is the vacuum permittivity and ε_r is the relative (dielectric) permittivity. Here, the real part of complex dielectric permittivity of gold is used. The dielectric permittivity of air as the incidence medium is denoted as ε_d. The electric displacement field derived from the Maxwell's equation is continuous across the interface. With the continuous normal component of D across the interface and the permittivity sign difference for metal and dielectric, the electric field changes direction passing through two different media. This characteristic will only be satisfied if there is a normal component for electric field across two regions that is TM polarization.

We are interested in metals with the large negative real part and a very small imaginary contribution to the dielectric constant for the design wavelength, such as gold. The SPP damping while propagating along the interface will be determined by the imaginary part of the wave vector parallel component. As the researchers demonstrated their results for noble metals, such as gold and silver in metal-dielectric interface, there will be a high field confinement at the interface while the losses remain minimum. When the plasmonic excitations occur, the left side of equation (5) matches the wave vector of the excited SPP (k_{spp}), that is the equivalence of the interaction of incident radiation and lth diffracted order with the wave vector of the SP at the interface.

The FDTD as a powerful engineering tool allows for the effective and powerful simulation and analysis of sub-micron devices with very fine structural details. The FDTD algorithm was originally proposed by K. S. Yee in 1966 [25]. In order to investigate the optical response of plasmonic-based MSM-PD, finite difference time-domain (FDTD) numerical method is used as a premier solution for the simulation of propagating electromagnetic field by solving Maxwell's curl equations in time domain. The computational mesh points (grids) are made up of unit cells, and the electric (E) and magnetic (H) fields are arranged at special places of the computational domain denoted by (i, j, k) with respect to Ampere and Faraday's laws. The FDTD method is able to model light propagation, scattering, diffraction, reflection, and polarization effects.

The Opti-FDTD software package developed by Optiwave Inc. was used to perform a 2D simulation, and it is the first software to employ the Lorentz–Drude model into the FDTD algorithm to calculate the transmission and reflection spectra. The FDTD simulation results have demonstrated significant light-capture performance through periodic 1-D slit arrays, which is useful for the design of ultrafast MSM-PDs.

In this model, the light wave hits on the nano-grating structures perpendicularly and passes through the subwavelength aperture and finally reaches to the semiconductor substrate.

This chapter's focus is on nano-gratings architectures that can increase the MSM-PD responsivity by exciting the SPPs and manipulation of light in the subwavelength slit. It has been reported that the rectangular nano-gratings produce the best absorption process, however, we will present some new more qualified nano-gratings with more efficient performance. The most influential parameters controlling the device light absorption can be categorized in two general groups. One can be the basic structural differences for nano-gratings clearly distinguishable with respect to their cross sections such as nano-gratings shape and dimensions, metal nano-grating heights, duty cycle, and number of nano-gratings on each side of the central slit. The other one is optimization process for subwavelength aperture region as subwavelength slit width, and electrodes (under layer) thickness. Furthermore the incident light polarization and angle of incidence play an important role to produce quality light absorption in optimized plasmonic-based MSM-PD. To verify the importance of these features, the power flow through the central subwavelength slit for plasmonic-based MSM-PD is compared with the amount of power reaching the active region for conventional MSM-PD. Improved light interaction process with different nano-grating shapes and geometries results in verification of the simulated results for the design and development of high responsivity MSM-PDs which have applications in high-speed optical fiber communication, high-speed sampling, and chip to chip interconnectors. The light absorption enhancement factor (LAEF) is introduced as a dimensionless quantity to measure the optimal absorbed radiation ratio to the whole incident power [26]. Therefore the impact of nano-gratings implementation on the quality of light flux transmitted into the active region of the MSM-PD is well approved compared with the conventional MSM-PD without the nano-gratings.

SIMULATION RESULTS AND DISCUSSION

Electric Field Distribution inside the Gaas Substrate

In this subsection, local electric field intensity distribution for conventional and plasmonic-based MSM-PD structures will be clarified through simulations to evaluate the adequacy of plasmonic-based structure. We use a custom designed Matlab code to show the density plot and the transmitted power into the substrate for two different structures. While the radiation reaches to the structure normally, Fig. 4(a) represents the electric field distribution in a conventional device without nano-gratings but with the Au contacts just on the GaAs substrate. The slit width is 100-nm and the under layer height is 60-nm. Therefore, in this situation, the light transmission inside the substrate is not influenced by the surface plasmon excitations and the incident light normally passes through the subwavelength aperture. Different colors represent each point's electric field strength in the density plots. Also, rectangular nano-grating structures are designed and deposited on the metal contacts to take the plasmonic effects into account. Fig. 4(b) demonstrates the electric field confinement in the GaAs substrate at cross section of plasmonic-based MSM photodetector device and the field concentration is due to SPP coupling with nano-structures. The maximum intensity appears for the part of the substrate located just under the slit [27].

Due to the plasmonic interactions and the confinement of light into the central subwavelength slit, it is obvious from Fig. 4 that the grating assisted MSM-PD tendency is to concentrate the power (or energy) into the photoactive region which is just below the central slit. Therefore, implementation of the nano-gratings enables the photodetector to lead the light into the central aperture quite effectively. Hence, it is important to obtain optimized geometrical parameters for efficient light confinement, and it is what we describe in the following subsections.

Effect of Duty Cycles on the Laef

For 2 similar nano-gratings with the discrepancy in profile shapes, the light harvesting ability can be different. The Au nano-grating profile and geometry on the GaAs substrate can make changes in the absorption enhancement spectrum. Therefore it is interesting to discuss the effects of duty cycle on the LAEF of the nano-structured MSM-PDs. Duty cycle (DC) of corrugations is the percentage of ridges width to the nano-grating period, i. e. 60% DC refers to the ratio of the nano-gratings grooves width to the ridges width of six to four in one period. We specify the optimum duty cycle for trapezoidal and rectangular-shaped nano-grating profiles which are designed with optimum heights [28].

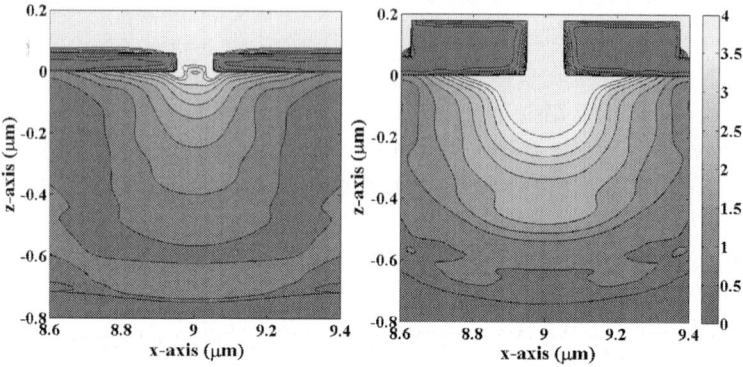

Figure 4: Field distribution at the cross section (a) conventional MSM-PD, (b) rectangular plasmonic-based MSM photodetector. The calculated total electric field intensity distribution inside the GaAs substrate is shown using the following parameters: the subwavelength slit width of 100-nm, gold under layer thickness of 60-nm, and the nano-gratings height of 100-nm for (b).

Under normal incidence, the LAEF spectra in rectangular-shaped nano-grating structures is calculated for different duty cycles, such as from 30% to 100%, as shown in Fig. 5, while the subwavelength aperture width has been kept constant at 50-nm, the under layer thickness and the nano-grating height were 20-nm and 120-nm, respectively. It can be inferred that the peak wavelength is different for each specific duty cycle and the maximum LAEF, with the amount of ~32.7, occurs for 40% duty cycle. It is clear that the duty cycle can affect the peak wavelength as well as the amount of light flux transmitted into the active area of the MSM-PDs.

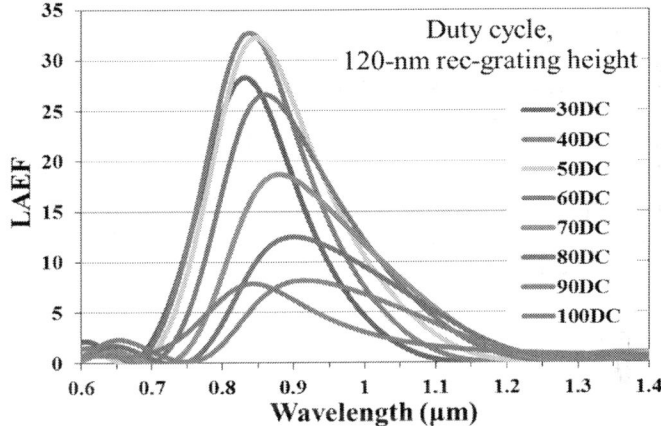

Figure 5: Light absorption enhancement factor spectra of MSM-PDs with rectangular-shaped nano-gratings for various duty cycles ranging from 30% to 100%.

The optimum duty cycle has also been specified for trapezoidal-shaped nano-grating profile with the subwavelength aperture width and under layer thickness of 50-nm and 20-nm, respectively. In this case, the maximum LAEF is obtained~31.5 for 40% DC which is the same duty cycle as the optimum DC for rectangular-shaped nano-gratings.

The results shown in Figs. 5 and 6 illustrate that the amount of LAEF is a function of duty cycle which grows gradually towards the 40% DC and drops down dramatically towards higher DCs and the peak wavelengths are also red-shifted for the LAEF curves of the higher duty cycles for both the structures. In the case of the rectangular and trapezoidal nano-grating profiles designed with their optimum nano-grating height, we can infer that the amount of light transmitted into the active region not only depends on the nano-gratings height and shape but also on the amount of duty cycles. Besides, for rectangular-shaped nano-grating profile shown in Fig. 5, 100% DC indicates that the whole structure is like a conventional MSM-PD having a thick under layer with height of about 140-nm, the resulting thickness is the sum of under layer thickness and the nano-grating height, while for trapezoidal shaped nano-gratings, there will be triangular grooves between the trapezoidal ridges for 100% DC, Fig. 7.

Impact of Subwavelength Aperture Width on the Laef

We discuss the impact of subwavelength aperture width on the light absorption and reflection for MSM-PDs and the results are discussed for different duty cycles. Compared with the incident light wavelength (λ_0), the aperture width is very small, hence only symmetric and fundamental SP modes will propagate into the slit. When the subwavelength slit width is much smaller than the incident light wavelength (λ_0), in addition to the light transmission and absorption caused by manipulation of metal nano-gratings, the light harvesting and confinement in the semiconductor substrate can be obtained via optimization of subwavelength aperture. Fig. 8 shows the simulation results of the absorption spectrum for several subwavelength aperture widths when the number of nano-gratings on each side of the slit (subwavelength aperture) is 9, and the nano-grating period and the nano-grating height are kept constant at 810-nm, and 100-nm, respectively. However, a portion of the lights is reflected in the central slit area, as shown in Fig. 9. The interesting point is that the range of wavelengths in the spectrum corresponding to the minimum reflection for the LRF curves in Fig. 9 is equivalent to the range of maximum LAEF for the same structure.

The results for less than 50-nm and more than 500-nm width slits are not presented here because the LAEF is reduced for very thin and very wide slits. The optimized slit width is selected as 50-nm which is much easier to fabricate compared with thinner slits and shows very promising results to improve the device performance.

The simulated results show that the LAEF decreases rapidly with the increase of the subwavelength aperture width from 500-nm to 50-nm. Figure 8 shows clearly that the LAEF is more than 12-times with 50% DC and about 13.5-times with 60% DC for a 50-nm subwavelength aperture width, the narrowest slit in this simulation, also with presentation of LAEF curves for 40% and 70% DCs of the aforementioned slit width, we show that 60% DC is optimized for this special structure. However, the LAEF is ~4-times with 50% DC but a little lower with 60 DC for a slit width of 250-nm and the LAEF wavelength is red shifted for all bigger DC.

The effective refractive index is a function of slit width for symmetrical SP modes when the slit experiences the TM incident wave. Therefore, with reducing the subwavelength aperture width, the effective refractive index increases and leads to an enhancement of the light absorption inside the active region.

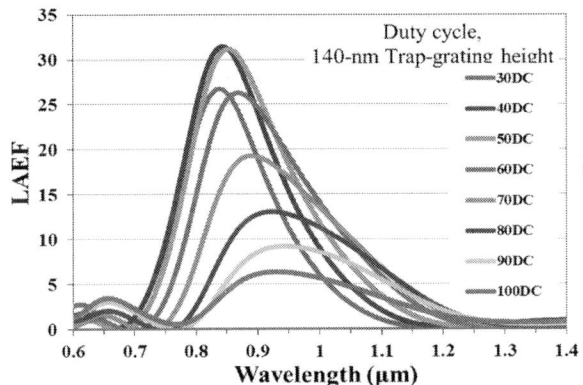

Figure 6: Light absorption enhancement factor spectra of MSM-PDs with trapezoidal-shaped nano-gratings for various duty cycles ranging from 30% to 100%.

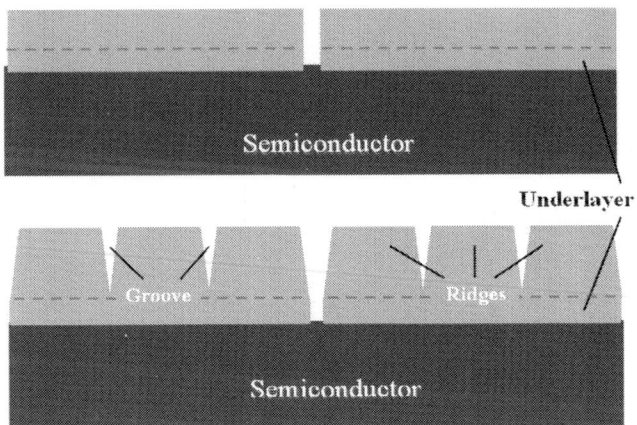

Figure 7: Cross-section of rectangular and trapezoidal-shaped nano-gratings while the duty cycle is 100%.

However, the idea for optimization of the MSM-PDs can be made more precisely by modifying the amount of transmitted light power through the subwavelength slit in the GaAs substrate [29]. Therefore, at normal incidence, we made this amendment by the subtractions of localized power near the top of the substrate and the power propagating to the bottom of the substrate, that is the LAEF values at the slit opening and beyond the slit opening at the outer edge of the substrate. The subtracted amount for LAEF is presented in Fig. 10 for the optimized geometry presented in Fig. 8, that is 50-nm slit width, 60% duty cycle. The maximum LAEF is almost unchanged and the fact that a minor amount of energy is lost from the substrate is justified as long as the device dimensions are optimized. Also the absolute values for Poynting vector in x direction are presented for 3 depths of the GaAs substrate, top, 0.1 µm from top and bottom. The $|S_x|$ value is negligible at the bottom of the substrate compared with the slit opening.

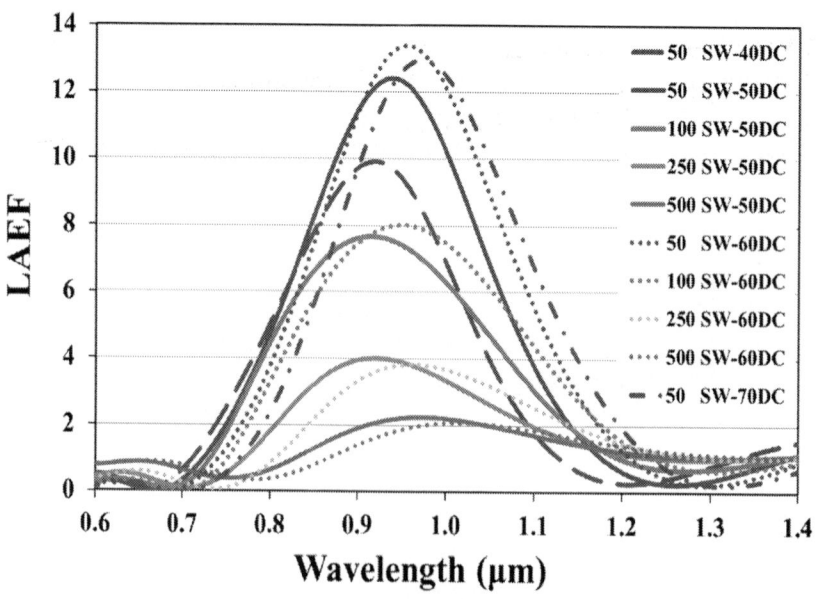

Figure 8: Light absorption enhancement factor spectra of MSM-PDs with rectangular-shaped nano-gratings. Different curves show the effect of slit width (SW) and duty cycle (DC) variations on the LAEF. Here, the under layer thickness and nano-gratings height are kept constant at 60-nm and 100-nm, respectively.

Impact of Incident Angle on Laef Curves

We present some results to investigate the effects of incidence angle upon the maximum LAEF for plasmonic-based MSM-PD device. The incident angle varies through a straight angle with negative and positive values, ranging from-90° to 90°, representing inclination from the normal incidence to left and right, respectively. While changing the angle of incidence for illuminated light in simulation, we show the device's most efficient light absorption enhancement for a specific angle. Fig. 11 shows maximum LAEF curve for various incident angles for the nano-grating structures with the subwavelength slit width of 50-nm and nano-grating height of 100-nm while the DC is kept constant at 60%, they are the parameters for the optimized curve producing the maximum LAEF in Fig. 8. The resonant wavelength is constant for most of the angles at 947-nm. The presented results indicate that the optimized incident angle is-46° for this geometry.

Nano-Grating Height Optimization in Plasmonic-Based Msm-Pd

With the variation of nano-gratings height, different sets of results show significant changes in the amount of light transmitted into the active area of the MSM-PD. Hence the height of the ridge is an effective parameter in optimization of the detector performance [30]. We present the simulation results for rectangular and trapezoidal MSM-PDs with different nano-grating heights. The incident light with TM polarization was perpendicularly illuminated on the groove profiles and we have calculated the amount of light flux transmitted into the slit for four different heights in rectangular and trapezoidal nano-grating assisted MSM-PDs.

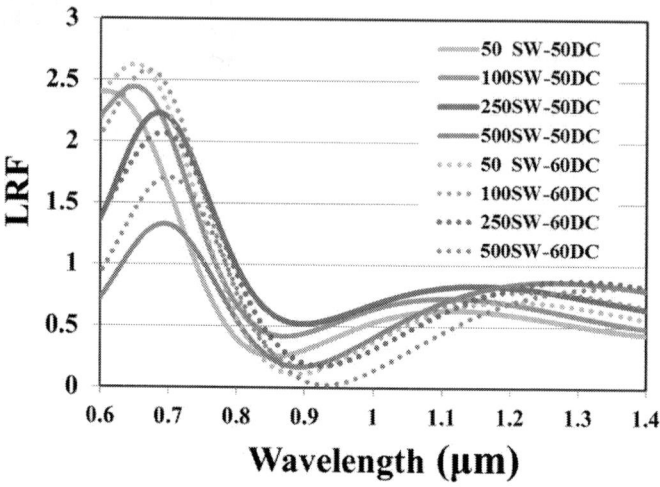

Figure 9: Light reflection factor spectra of MSM-PDs with rectangular-shaped nano-gratings. Different curves show the effect of slit width (SW) and duty cycle (DC) variations on the Light reflection factor.

Figure 12 shows the LAEF spectrum for different heights of the rectangular shaped nano-gratings in plasmonic-based MSM-PD, such as 80, 100, 120, 140-nm. Simulation results confirm that 120-nm is the optimized height for this design. The peak wavelength behaves like a sinusoidal manner and wavelength (λ) is red shifted as the ridge's height increases. The duty cycle is 60% while the subwavelength aperture width, and subwavelength aperture thickness are kept constant at 50-nm, and 20-nm, respectively. There are some interpretations to analyze the curves. The SPPs coupling process and the following expected absorption can easily occur for higher gratings and reduces after certain heights. This light absorption has a maximum for a specific wavelength in the spectra which varies in different heights, and for the wavelengths higher than the peak, the amount of LAEF decreases because the incident light might be coupled into radiative SPs rather than bound SP modes.

Also, several sets of numerical analysis are carried out to illustrate the effect of trapezoidal-shaped nano-gratings on the optimized height at which the maximum resonance transmission occurs. The simulations are performed for 4 different nano-grating heights of 100, 120, 140, and 160-nm with the subwavelength aperture width of 50-nm and the under layer thickness of 20-nm. The results are shown in Fig. 13. The

LAEF is 26.3 times in its maximum for the curve representing the 140-nm nano-grating height for the duty cycle of 60%.

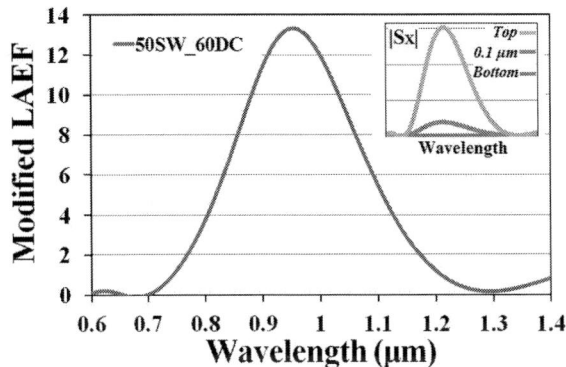

Figure 10: Modified LAEF for plasmonic-based MSM-PDs with 50-nm slit width, 100-nm nano-grating height, and 60% duty cycle. The inset represents the absolute value of Sx in different depths of GaAs substrate, the top, 0.1 µm depth from the substrate surface, and the bottom.

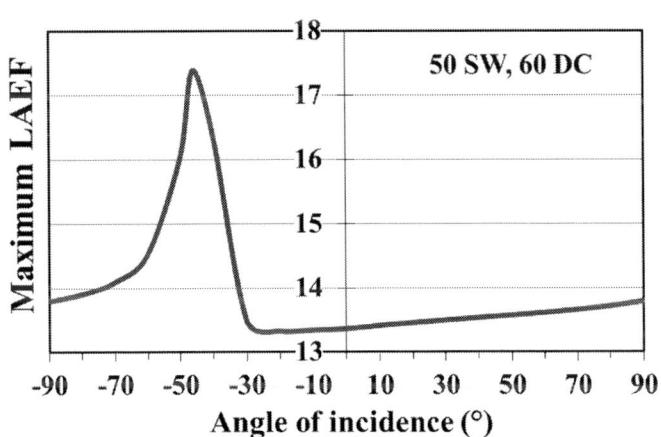

Figure 11: Maximum LAEF versus incidence angle characteristics for MSM-PDs with 50-nm slit width and 60% DC. Here, the subwavelength aperture height and nano-gratings height are kept constant at 60-nm and 100-nm, respectively.

From Figs. 12 and 13, it can be recognized that the optimum wavelength is red shifted for higher nano-gratings in both figures. However, trapezoidal nano-gratings height at maximum light absorption, 140-nm, is higher than the optimum height of its rectangular counterpart, 120-nm, while its resonant wavelength is blue shifted.

Nano-Grating Geometries Effect in Plasmonic-Based Msm-Pd

There is no doubt about the effects of nano-grating textured structures and geometries on light trapping inside the device active region as they are responsible for the creation of the SPPs which can assist for the light confinement in the subwavelength regions. Hence we analyze the MSM-PD device performance and its enhanced responsivity for different nano-grating shapes.

By involving plasmonics (i.e., metallic nano-gratings), the device performance has been improved due to the advances in nano-technology fabrication methods. Focused ion beam (FIB) lithography, electron beam (E-beam) lithography, and nano-lithography are the new approaches to fabricate nano-scale devices. These techniques can be used to obtain the state-of-the-art for very small nano-structures in order of tens of nano-meters which are the accepted dimension for visible and near infrared regions [31].

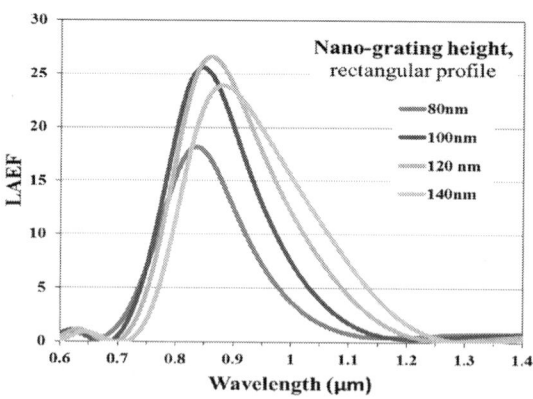

Figure 12: Light absorption enhancement factor spectra of the rectangular nano-gratings with different heights. Here, DC is 60% and subwavelength aperture width is 50-nm.

In general, a good light absorption performance is achieved for rectangular-shaped nano-gratings. It has been proved that the rectangular nano-gratings are the best option for the design of nano-gratings to improve the MSM-PDs performance. The normal wall nano-gratings are easily fabricated with lithography and etching, while in focused-ion beam (FIB) milling, the nano-gratings are rather taper than the rectangular one. The observation of scanning electron microscopy images represents the nano-gratings with taper walls rather than the rectangular nano-gratings. The metallic nano-gratings fabrication is not an easy process but the FIB lithography is an approved technique to develop elaborate structures with the nano-gratings. This nano-scale fabrication technology is very promising to access near field excitations and detection of the plasmon polaritons at air/Au interface.

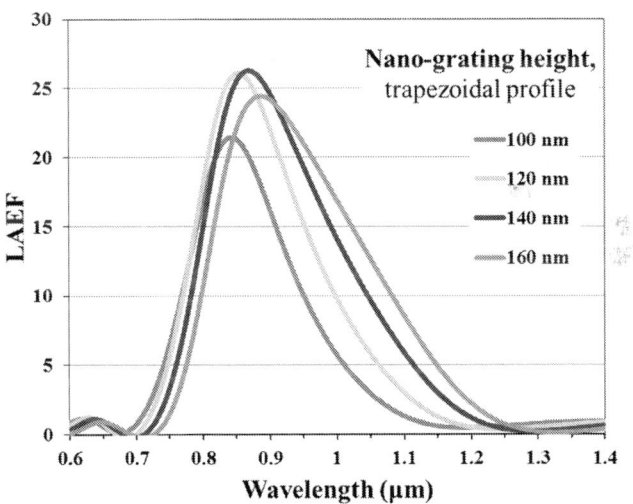

Figure 13: Light absorption enhancement factor spectra of the trapezoidal profile nano-gratings with different heights. Here, DC is 60% and subwavelength aperture width is 50-nm.

Here the plasmonic-based MSM-PD structures are set with the rectangular, trapezoidal, and ellipse-wall nano-grating profiles and their behaviors are compared with each other. It has been reported that the straight wall nano-gratings produce an optimum light absorption for plasmonic-based MSM-PDs, but under practical device manufacturing situations, the rectangular shaped profile is closer in appearance to semi-trapezoidal nano-gratings. Then we introduce and characterize grating-assisted MSM-PD which utilizes ellipse-wall nano-gratings and prove that it achieves better efficiency than rectangular and trapezoidal counterpart.

Metallic nano-grating geometries affect the LAEF in MSM-PDs. Figure 14 shows the light absorption enhancement spectra evolution for MSM-PDs with triangular, rectangular, trapezoidal with 0.4, 0.5, 0.8, and 0.9 aspect ratios, and the ellipse-wall nano-gratings with 0.5, and 0.9 aspect ratios. Aspect ratio (AR) is introduced to define a relationship between the width of upper and lower bases for taper and ellipse wall structures as a dimensionless coefficient smaller than unity. For ordinary structures, the lower base and for inverted ones the upper base is always bigger. These groove shapes along with the differences between their aspect ratios are shown in Fig. 15. The simulation results illustrate a strong confinement for the rectangular shaped structures, although the realistic subwavelength nano-gratings do not have the normal walls. The maximum LAEF for taper and rectangular profiles are quite close.

However, we introduce a novel grating structure for our design with ellipse walls which has recently been stated to have more light capturing efficiency than the rectangular and taper nano-grating profiles [32]. As stated the experimental results in [31], the analysis of atomic force microscope (AFM) systems and scanning electron microscope (SEM) images demonstrated the trapezoidal structures rather with curved walls than the linear walls.

Therefore, we have designed the ellipse-wall nano-gratings with legs satisfying the exponential equation having an exponential coefficient of 0.5. The exponential coefficient is supposed to satisfy the exponential function of $z=C\,e^x$ for nano-gratings lateral walls, where C is the exponential coefficient. Performing the simulations on these ellipse-wall nano-structures gives a better view of more realistic condition for nano-scale devices. The simulation is performed for the under layer thickness of 20-nm, the nano-grating height of 120-nm, and the subwavelength aperture width of 50-nm which are optimized values for a MSM-PD with rectangular symmetric nano-gratings [33].

Besides, the results of light absorption for trapezoidal structures with different aspect ratios are presented. In the case of slanted walls, the increase of slit opening width is a drawback for reflection of gap plasmon from the upper termination resulting in a weak LAEF, because the cavity nature which is responsible for resonant absorption faded away with variation of the slit opening width. However, the decrease of taper aspect ratio leads to a blue shift resonance position. Figure 14 shows this slight blue shift while this parameter changes from 0.9 to 0.4 for tapered nano-grating structures.

As shown in Fig. 14, the plasmonic interactions are more efficient for ellipse wall nano-gratings. We know that the localization of optical energy around sharp corners is remarkable. The non-linear design of ellipse-wall nano-gratings enables the possibility to improve energy concentration in the active region of the MSM-PD device because the energy flow is facilitated through the interface in comparison with the rectangular nano-gratings. Depending on the aspect ratio parameter, that is the normalized value of bottom side to the top side width as shown in Fig. 15, nano-gratings can increase their LAEF. Ellipse-wall nano-gratings improve their performance from LAEF of 20 (for 0.5 aspect ratio) to 28 (for 0.9 aspect ratio). Even the maximum LAEF of rectangular nano-grating, 26.5, is less than 0.9 aspect ratio ellipse-wall structures. In addition, the maximum peak wavelength is red shifted for rectangular structure. While the 0.9 aspect ratio ellipse-wall nano-grating is the most suitable structure in light absorption, it is worth noting that the 0.5 aspect ratio ellipse-wall nano-gratings offer a better transmission in comparison to their trapezoidal counterparts. They almost doubled their efficiency compared with the 0.5 aspect ratio taper profiles from 10.5 to 20.3.

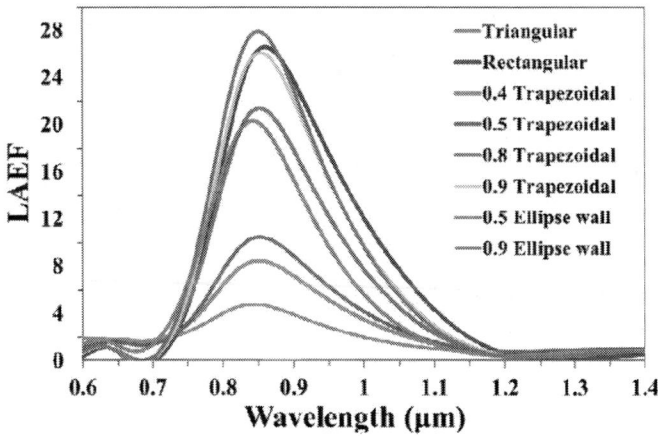

Figure 14: Light absorption enhancement factor spectra of MSM-PD with triangular, rectangular, taper nano-grating structures with 0.4, 0.5, 0.8, and 0.9

aspect ratios, and ellipse wall nano-gratings with 0.5 and 0.9 aspect ratio and exponential coefficient of+5. The duty cycle of corrugations, subwavelength aperture thickness, and the nano-gratings thickness are 60%, 20-nm, and 120-nm, respectively.

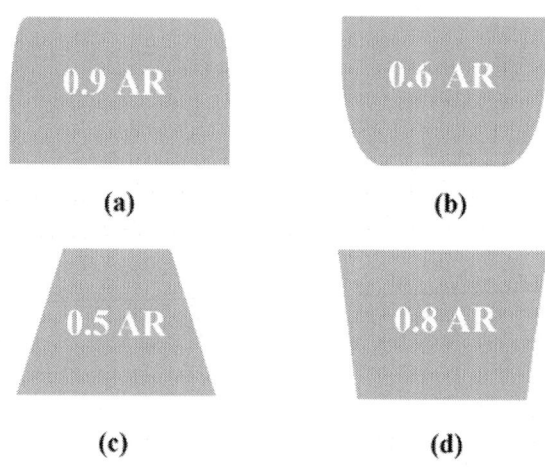

(a)

(b)

(c)

(d)

Figure 15: Ellipse-wall nano-gratings with the aspect ratio (AR) value of 0.9 and 0.6 for (a) and (b), respectively. For (a) and (b), the lateral walls are function of exponential equation with the exponential coefficient of+5 for (a) and-5 for (b). (c) and (d) are taper (trapezoidal) and inverted trapezoidal structures with 0.5 and 0.8 aspect ratios. For inverted structure, the lower base is bigger. The aspect ratio defines a relationship between two parallel bases widths of trapezoidal and ellipse-wall nano-gratings.

Impact of Under Layer Thickness

Here, we discuss the impact of under layer or subwavelength aperture thickness (Uth) on the amount of transmitted light through the active region of the MSM-PD. Under layers are the photodetector's metallic electrodes which metallic nano-gratings can be designed on them. Optimization of this parameter also affects the light absorption enhancement in the MSM-PDs. The thick under layers absorb a proper amount of the incident lights and the carrier recombination become more prominent which reduces the internal quantum efficiency. The thinner under layer leads to better electrical properties moreover they have advantages of material saving and higher carrier collection efficiency; however they lose light harvesting proficiency which can be resolved by implementation of nano-grating structures of the same metal. The reduction of the under layer thickness enhances the LAEF, because in this case the EOT can also occur in the flanked slits other than the central aperture. Here, the under layer thickness attenuation effect is shown, which assists the light absorption in the device in addition to the light absorption enhancement caused by optimization of other parameters, i.e. nano-gratings shape. As expected, an ideal rectangular nano-grating shape with the thinnest under layer contributes more effectively in the light absorption in comparison to the thicker under layer heights. However, the amount of LAEF for ellipse wall structure is higher than the rectangular grooves, because in this design, the central slit opening has no corner which avoids useless light confinement at sharp edges, also the slit opening is broader, so there is less reflection for the illuminated light at the top due to the grooves' non-linear walls and a greater part of the energy involves in the SPP coupling at metal-air interface. Design of the nano-gratings with the thickness of 120-nm, which is the optimized height for rectangular nano-grating as shown in Fig. 12, results in about 32.8 fold enhancement for ellipse-wall nano-gratings with 0.9 aspect ratio and exponential coefficient of 0.5 when the duty cycle is fixed at 40% and the subwavelength slit width is 50-nm, but ellipse-wall nano-gratings quite close absorption peak to rectangular nano-grating with the height of 120-nm is the result of doing simulations for optimized rectangular nano-grating.

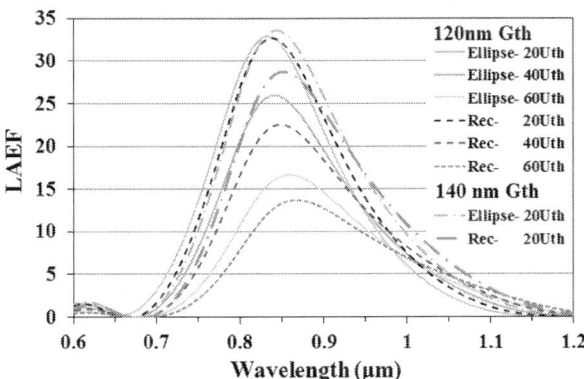

Figure 16: Light absorption enhancement factor (LAEF) calculation of different under layer thicknesses (Uth) for the rectangular and 0.9 aspect ratio

ellipse-wall nano-grating structures with exponential coefficient of+5 for two different nano-grating thicknesses (Gth).

While absorption is enhanced for thinner under layers, useful results will be revealed through comparison of absorption curves for optimized nano-grating thickness of ellipse-wall nano-gratings and rectangular nano-gratings, 140-nm and 120-nm, respectively. Our simulated results indicate better enhancement of LAEF for ellipse-wall nano-gratings with optimized thickness of 140-nm that is 33.6 compared with maximum LAEF of 120-nm rectangular nano-grating which is 32.7. These results are shown in Fig. 16.

In addition to the influence of under layer thickness on the amount of transmitted light flux into the device active region, we can also demonstrate under layer's direct influence on other structural parameters like duty cycle. Looking back at sections 5.2 and 5.3, we notice the LAEF curves calculated in different duty cycles. In Fig. 8, for the slit widths of 50-nm, the 60% is the optimized duty cycle while in Fig. 5 and 6, the optimized duty cycle is 40%. We mentioned this to prove the fact that the optimized duty cycle varies for the structure with different under layer thickness, the under layer thickness of 20-nm and 60-nm in Fig. 5 and Fig. 8, respectively.

CONCLUSIONS

We discussed the interaction of illuminated light as electromagnetic waves through the central sub-wavelength slit on a metallic thin film surrounded by periodic nano-gratings. The concept of SPPs has been introduced and the excited SPPs are generated at metal-dielectric interface which are used for plasmonic-based applications. Plasmonics offers the ability to concentrate light into subwavelength volumes in ultra-small optoelectronic devices utilizing high-speed and broad bandwidth. Plasmonics has also made impression in nano-scale photodetector development. Photodetectors play key role in development of modern optical communication technology. Surface plasmon resonances have found practical applications in sensitive photodetectors recently. We have analyzed the performance and advantages of plasmonic-based MSM photodetectors and modeled their light absorption enhancement. The FDTD simulation tool has been used to analyze and optimize the impact of the physical and geometrical parameters on the amount of transmitted light into the MSM-PD structures. In order to maximize the flow of energy into the device active region, the performance of different nano-grating profiles and their effects on the efficiency of the photodetector have be evaluated using Drude-Lorentz dielectric function via FDTD algorithm. The corrugations on the surface lead to an effective impedance for the surface modes which favors to the resonant coupling of the SPs with incident electromagnetic wave and hence facilitates the enhancement of light transmission. The main motivation to design the nano-gratings on the MSM-PD electrodes is to assist and improve the light transmission into the slit. With the grating assisted MSM-PDs, the device could benefit the small spacing between the two top contacts, namely the central slit, for a fast response of optical pulses. The simulated results reveal that the amount of LAEF is much better than the conventional MSM-PDs. A substantial light absorption enhancement has been found for the symmetrical MSM-PD devices. Two distinct mechanisms targeting the absorption enhancement in MSM-PDs are namely, the metal nano-gratings assisted light absorption and the subwavelength slit Fabry-Perot resonances. Feasibility of developing MSM photodetectors with high-responsivity and high-speed characteristics has made them reliable choices for high-speed optical communication systems. We have studied the transmission of TM-polarized light through subwavelength apertures in metallic films flanked by metallic nano-gratings in plasmonic-based MSM-PDs and optimized the interdigitated electrodes thicknesses and the nano-grating shapes. Simulation results confirm that the plasmonic–based grating–assisted MSM-PD is more efficient in light absorption compared with a conventional MSM-PD. The light energy confinement in the nano-scale and the produced focal point of the incident beam in plasmonic device compared with conventional MSM-PD confirm the characteristics of a nano-plasmonic lens. The demonstration of device optimization results assists to improve the concept of novel high responsivity, plasmonic-based MSM-PDs for high speed applications in optical communication. These results provide useful information for the design and fabrication of nano-scale optoelectronic devices.

REFERENCES

1. G. P. Wiederrecht. Handbook of nanoscale optics and electronics, Elsevier, Amsterdam, First edition 2010. ISBN: 978-0-12-375178-2.

2. C. DeCusatis, and I. Kaminow. The Optical Communications Reference, Academic Press 2009. ISBN: 978-0-12-375163-8.

3. H. Zimmermann. Silicon Optoelectronic Integrated Circuits, Springer Berlin Heidelberg 2004; 13, 1-23, doi: 10.1007/978-3-662-09904-9_1.

4. B. P. Pal. Fundamentals of Fibre Optics in Telecommunication and Sensor Systems, bohem press, 1992.

5. G. P. Agrawal. Fiber-Optic Communication Systems, John Wiley & Sons, Inc., Fourth Edition, 2002. doi: 10.1002/9780470918524.

6. S. Y. Chou, Y. Liu, and P. B. Fischer. Tera-hertz GaAs metal-semiconductor-metal photodetectors with nanoscale finger spacing and width, Electron Devices Meeting, 1991. IEDM '91. Technical Digest., International, 745-748. doi: 10.1109/IEDM.1991.235316.

7. W. Zhang, A. K. Azad, and J. Han. Resonant Excitation of Terahertz Surface Plasmons in Subwavelength Metal Holes, Hindawi Publishing Corporation, Active and Passive Electronic Components, 2007, Article ID 40249, 8 pages, doi:10.1155/2007/40249.

8. R. Umeda, C. Totsuji, K. Tsuruta, and H. Totsuji. Dispersion Models and Electromagnetic FDTD Analyses of Nanostructured Metamaterials using Parallel Computer, Memoirs of the Faculty of Engineering, Okayama University January 2009; 43, p. 8.

9. A. D. Rakic, A. B. Djurišic, J. M. Elazar, and M. L. Majewski. Optical Properties of Metallic Films for Vertical-Cavity Optoelectronic Devices, Applied Optics 1998; 37(22), 5271-5283. http://dx.doi.org/10.1364/AO.37.005271.

10. M. Bordovsky et al. Waveguide design, modeling, and optimization: from photonic nanodevices to integrated photonic circuits, in Proceedings of the SPIE 5355, Integrated Optics: Devices, Materials, and Technologies VIII, 65, doi: 10.1117/12.526976, May 28, 2004.

11. B. Ung. Study of the interaction of surface waves with a metallic nano-slit via the finite-difference time-domain method, M.Sc. Thesis, Laval University, Quebec, Canada, Ch. 3, 2007. http://theses.ulaval.ca/archimede/fichiers/24879/24879.html.

12. F. F. Masouleh, N. Das, H. Mashayekhi. Impact of duty cycle and nano-grating height on the light absorption of plasmonics-based MSM photodetectors, in Proceedings of the 12th IEEE Int. Conf on Numerical Simulation of Optoelectronic Devices 2012; Shanghai, China, 13-14. doi:10.1109/NUSOD.2012.6316521.

13. T. W. Ebbesen, H. J. Lezec, H. F. Ghaemi, T. Thio, P. A. Wolff. Extraordinary optical transmission through sub-wavelength hole arrays, Nature 1998; 391, 667-669. doi:10.1038/35570.

14. H. J. Lezec, A. Degiron, E. Devaux, R. A. Linke, L. Martin-Moreno, F. J. Garcia-Vidal, T. W. Ebbesen. Beaming light from a subwavelength aperture, Science 2002; 297, 820-822. doi: 10.1126/science.1071895.

15. Y. Ding, J. Yoon, M. H. Javed, S. H. Song, R. Magnusson. Mapping surface-plasmon polaritons and cavity modes in extraordinary optical transmission, IEEE Photonics Journal 2011, 3, 365-374. doi:10.1109/JPHOT.2011.2138122.

16. S. A. Maier. Plasmonics: Fundamentals and Applications, Springer, 2007; ISBN 978-0-387-37825-1.

17. G. T. Reed and A. P. Knight. Silicon photonics: an introduction, John Wiley and Sons 2004; ISBN 0-470-87034-6.

18. S. Collin, F. Pardo, R. Teissier, J. L. Pelouard. Horizontal and vertical surface resonances in transmission metallic gratings, Journal of Optics A: Pure and Applied Optics 2002; 4, 154-160. doi:10.1088/1464-4258/4/5/364.

19. D. de Ceglia, M. A. Vincenti, M. Scalora, N. Akozbek and M. J. Bloemer. Enhancement and inhibition of transmission from metal gratings: Engineering the Spectral Response, at http://arxiv.org/abs/1006.3841, 2010.

20. J. A. Porto, F. J. García-Vidal and J. B. Pendry. Transmission resonances on metallic gratings with very narrow slits, Physical Review Letters 1999; 83, 2845-2848. doi: http://dx.doi.org/10.1103/PhysRevLett.83.2845.

21. A. Barbara, P. Quemerais, E. Bustarret and T. Lopez-Rios. Optical transmission through subwavelength metallic gratings, Physical Review B 2002; 66, Article ID 161403. doi: http://dx.doi.org/10.1103/PhysRevB.66.161403.

22. H. Raether. Surface Plasmons on Smooth and Rough Surfaces and on Gratings, Springer-Verlag, Berlin, 1988. doi:10.1007/BFb0048323.

23. L. Martín-Moreno, F. J. García-Vidal, H. J. Lezec, A. Degiron, and T. W. Ebbesen. Theory of highly directional emission from a single subwavelength aperture surrounded by surface corrugations, Physical Review Letters 2003; 90, Article ID 167401. doi: 10.1103/PhysRevLett.90.167401.

24. J.A. Shackleford, R. Grote, M. Currie, J.E. Spanier, B. Nabet. Integrated plasmonic lens photodetector Appl. Phys. Lett. 2009; 94, 083501.http://dx.doi.org/10.1063/1.3086898.

25. K. S. Yee, Numerical solution of initial boundary value problems involving maxwell's equations in isotropic media, Antennas and Propagation, IEEE Transactions on 1966; 14(3), 302-307. doi:10.1109/TAP.1966.1138693.

26. E. Chen, and Y. S. Chou. Polarimetry of thin metal transmission gratings in the resonance region and its impact on the response of metal-semiconductor-metal photodetectors, Applied physics letters 1997; 70(20), 2673-2675. http://dx.doi.org/10.1063/1.118990.

27. F. F. Masouleh, N. K. Das, and H. R. Mashayekhi. Assessment of amplifying effects of ridges spacing and height on nano-structured MSM photo-detectors, Journal of Optical and Quantum Electronics April 2014; 46(4). doi:10.1007/s11082-014-9900-8.

28. N. Das, F. F. Masouleh, and H. R. Mashayekhi. A Comprehensive Analysis of Plasmonics-Based GaAs MSM-Photodetector for High Bandwidth-Product Responsivity, Advances in OptoElectronics, 2013, Article ID 793253, 10 pages, 2013. doi:10.1155/2013/793253.

29. N. K. Das, F. F. Masouleh, and H. R. Mashayekhi. Light Absorption and Reflection in Nano-Structured GaAs Metal-Semiconductor-Metal Photo-Detectors, IEEE Transactions on Nanotechnology Sep. 2014; 13(5), 1-8, doi: 10.1109/TNANO.2014.2336857.

30. F. F. Masouleh, N. K. Das, and H. R. Mashayekhi. Optimization of light transmission efficiency for nano-grating assisted MSM-PDs by varying physical parameters, The Journal of Photonics and Nanostructures-Fundamentals and Applications February 2014; 12(1), 45-53. doi: 10.1016/j.photonics.2013.07.011.

31. N. Das, A. Karar, M. Vasiliev, C.L. Tan, K. Alameh, Y.T. Lee. Analysis of nano-grating-assisted light absorption enhancement in metal–semiconductor–metal photodetectors patterned using focused ion-beam lithography, Optics Communications 2011; 284(6), 1694–1700. doi: 10.1016/j.optcom.2010.11.065.

32. Y. Liang, W. Peng, R. Hu, and H. Zou. Extraordinary optical transmission based on subwavelength metallic grating with ellipse walls, Optics Express 2013; 21(5), 6139-6152. doi: 10.1364/OE.21.006139.

33. F. F. Masouleh, N. K. Das, and H. R. Mashayekhi. Comparison of different plasmonic nano-grating profiles for quality light absorption in nano-structured MSM photo-detectors, Optical Engineering 2013; 52(12), 127101. doi:10.1117/1.OE.52.12.127101.

Experimental Study of Fiber Laser Cavity Losses to Generate a Dual-Wavelength Laser Using a Sagnac Loop Mirror Based on High Birefringence Fiber

Manuel Durán-Sánchez[1], R. Iván Álvarez-Tamayo[2],
Evgeny A. Kuzin[3], Baldemar Ibarra-Escamilla[3], Andrés
González-García[3], and Olivier Pottiez[4]

[1]Mecatrónica, Universidad Tecnológica de Puebla (UTP), Puebla, México

[2]Facultad de ciencias físico-matemáticas Benemérita Universidad, Autónoma de Puebla (BUAP), Puebla, México

[3]Departamento de Óptica, Instituto Nacional de Astrofísica, Óptica y Electrónica (INAOE), Puebla, México

[4]Departamento de Fibras Ópticas, Centro de Investigaciones, en Óptica (CIO), León, Guanajuato, México

INTRODUCTION

Dual wavelength fiber lasers (DWFL) research has increased considerably in recent years due to the potential applications of these optical devices in diverse investigation areas. Interest of use of DWFL includes areas such as fiber sensors, wavelength division multiplexing, optical communications systems, optical instrumentation and recently in microwaves generation [1-4], among others.

DWFL are considered profitable optical sources because of their advantages such as low cost, easy and affordable optical structures, low losses insertion and space optimization. Principal issue to generate two simultaneous laser lines resides in the cavity losses adjustment. In DWFL designed with Erbium-doped fiber (EDF) as a gain medium there is a strong competition between the generated laser lines due to the EDF's homogeneous gain medium behavior at room temperature. To reduce the competition between the wavelengths, several techniques have been reported aiming to achieve stable multi-wavelength laser oscillations [5-8].

Moreover, fiber Bragg gratings (FBG) have been extensively used in DWFL cavities design due to their advantages as optical devices including easy manufacture, fiber compatibility, low cost and wavelength selection among others. FBG's wavelength selection property is commonly used as a narrow band reflector inside the laser cavity to generate a laser line at a specific wavelength. Several DWFL experimental setups using FBG's have been reported including use of a FBG written in a high birefringence or in a multimode fiber [6-11].

In a large majority of DWFL using EDF and FBG's, the laser cavity losses correspond to different generated laser lines at a specific wavelength position over the gain medium spectrum. The generated wavelength should be balanced to achieve two simultaneous laser lines. Consequently, both oscillation lines have the same pump threshold. Commonly the wavelengths adjustment is realized through arbitrary methods as use of polarization controllers (PC) and variable optical attenuators (VOA) [7, 12,13]. With the progress on DWFL research studies have been followed two different pathways in order to enhance stability of the simultaneously generated laser lines by improving the cavity losses adjustment methods.

On the one hand, the research focuses on incorporating of cutting-edge devices in an effort to obtain more stable and efficient dual laser emissions. In such a way that these researching works reports the use of newly developed optical fibers such as photonic crystal fibers, leading to use optical devices that allow the exploit of nonlinear optics [14-16]. Most of the reported works on this area tend to have more complex designs and non-straightforward settings. By the other hand, a second pathway is in function of simplicity and optimization of laser cavity length, taking into account that a reduced cavity length implies a decrease of laser modes within the cavity, allowing, in a first instance analysis, a dual laser emission with lower instability, a simple adjustment of the competition between laser lines with a substantial reduction of implementation space that can improve the results repeatability [17, 18].

In recent years, obtaining of dual-wavelength laser emission does not represent an advance by itself in DWFL progress because the increasing need to analyze the behavior of the competition between the generated laser lines obtained by the cavity losses adjustment methods. Using arbitrary methods like adjustment by polarization controllers and variable optical attenuators do not allow a behavioral analysis of the competition between generated wavelengths because these methods do not have a measurable physical variable to characterize the adjustment and difficultly can provide repeatability in results.

The spectral selectivity of the interferometer is caused by birefringence that has to be introduced to the loop. A lot of effort has been made to suggest and investigate a variety of FOLM designs. Ma et al. [20] demonstrated polarization independence of the Hi-Bi FOLM. Liu et al. [21] reported a study of an optical filter consisting of two concatenated Hi-Bi FOLMs. Lim et al. [22] analyzed the behavior of an FOLM with a fiber loop consisting of two Hi-Bi fibers connected in series. The transmittance spectrum of the FOLM presents a periodic behavior with maxima and minima depending on the Hi-Bi fiber length and birefringence. For dual-wavelength lasers, low contrast offers the advantage of smoother cavity loss adjustment for the generated wavelengths where the principal mechanism of the adjustment of the cavity loss is the shift of the wavelength of the reflection maxima of the FOLM. The wavelength shift is achieved by the change of the temperature of the Hi-Bi fiber. This method allows generating two wavelengths with a well-controlled ratio between their powers [19].

Moreover, the tuning of the laser generated wavelengths promises to be an advantage for DWFL microwave generation application making it possible through the tuning of separation between wavelengths. A simple method of wavelength tuning is related to the Bragg period modification of a FBG. Wavelength tunable DWFL were reported [16-19]. In most configurations the FBG's are used with Bragg wavelength shift by temperature change [23], compression or stretch [18, 24]. Most of the techniques reported before as a matter of fact realize an adjustment of the losses between the two wavelengths to achieve stable dual-wavelength generation. In spite of the numerous papers reporting dual-wavelength generation, to the best of our knowledge no investigations were reported on the relation between the losses for generated wavelengths that enables simultaneous dual-wavelength generation.

M. A. Mirza [25] in 2008 presented the theoretical and experimental analysis of the design of a Sagnac loop filter (SLF) with periodic output spectrum controlled by cascading a small birefringence loop (SBL) with a high birefringence loop (HBL) with a tuning of the amplitude and wavelength of the spectrum of the filter through mechanical rotation. In this work is mentioned that the proposed design may have potential application in the design of Erbium-doped fiber lasers for multiple wavelengths generation in the C band and also can be used as a tuning tool for competition between the generated laser lines.

H. B. Sun [26] published in 2010 a DWFL with wide tuning based on a Hi-Bi FOLM and the use of polarization controllers inside the loop for adjustment of the loss within the ring cavity proposed. The laser wavelength can be tuned flexibly within the range of 1525 nm to 1575 nm by adjusting the polarization controller. The separation between the two generated wavelengths is adjustable by changing the length of the Hi-Bi fiber of the FOLM loop. Also proves the modes stability of the two laser lines at room temperature with a variation of the peak output power of about 0.5 dB over 40 minutes of operation.

K. J. Zhou [27] in 2012 reported the use of an all-PM Sagnac loop periodic filter as a frequency selector in a Erbium-doped fiber ring laser. The laser with a 1 nm interval filter generates four simultaneous and stable wavelengths with equal frequency spacing to overcome the homogeneous broadening of Erbium-doped fiber as a gain medium at room temperature. Polarizer controllers are used inside the ring cavity to adjust the laser lines emissions. The experiment confirm that this kind of filter should be robust to environmental changes.

This chapter proposes the application of a Sagnac fiber optical loop mirror with a high-birefringence fiber on the loop (Hi-Bi FOLM) used as a spectral filter to adjust finely the laser cavity losses, reducing the competition between generated laser wavelengths by temperature variations on the FOLM fiber loop. This control allows characterizing the competition behavior with temperature variations to achieve a better adjustment to obtain dual-wavelength laser emission. The appropriate choice of the angles of both ends of the Hi-Bi fiber allows a reflection minimum between 0 and 0.9 without substantial wavelength shift. The reflection maximum is always equal to 1 [19].

In this chapter the application of an all-fiber Hi-Bi FOLM to balance the losses within a dual-wavelength fiber laser is presented. An analysis of the losses is performed by charactering the FBG's reflections over the transmission spectrum of the FOLM when the laser wavelengths are generated, allowing the study of the fine adjustment of the FOLM transmission spectrum wavelength shift by temperature variation in the Hi-Bi fiber loop of the FOLM necessary to achieve dual-wavelength laser emission.

NUMERICAL ANALYSIS OF SAGNAC HI-BI FOLM FOR DUAL-WAVELENGTH LASER APPLICATION

Numerically analysis for variation of the transmission spectrum of a Hi-Bi FOLM with the twist of the fiber in the loop can be an important tool for dual-wavelength fiber lasers design. The Hi-Bi FOLM shown in Figure 1 consists of a fiber coupler with a coupling ratio of $\alpha/1-\alpha$, which is assumed to be independent of wavelength. The output ports (3 and 4) are fusion spliced to a Hi-Bi fiber with arbitrary angles between the axes of the Hi-Bi fiber and the axes of the coupler ports. The segments where the Hi-Bi fiber is spliced to the coupler ports are placed on rotation stages. The Hi-Bi fiber is placed on a thermoelectric cooler to shift the wavelength dependence of the filter transmission. A light beam with electric field E_i enters through port 1; the transmitted beam with electric field E_T exits from port 2.

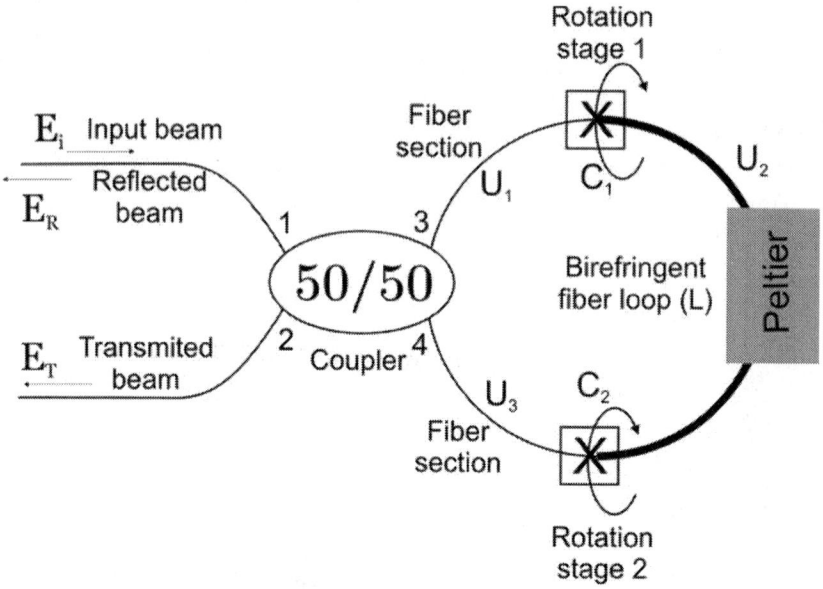

Figure 1: High birefringence fiber optical loop mirror.

To calculate the transmission of the FOLM, we used the approach developed by Mortimore [28]. For a single input field E_i, a transmitted field E_T is given by:

$$ET=(ETxETy)=((2\alpha-1)Jxx(1-\alpha)Jxy+\alpha Jyx-\alpha Jxy-(1-\alpha)Jyx(1-2\alpha)Jxx)$$
$$(EixEiy),$$

(1)

where the J matrix is calculated as the product of matrices corresponding to all elements in the loop:

$$J=U_1 \cdot C_1 \cdot U_2 \cdot C_2 \cdot U_3,$$

(2)

where matrices U_1 and U_3 represent the coupler ports; the matrices C_1 and C_2 represent the coordinate rotation accounting for the angles

dual wavelength laser stability. This is accomplished by adjusting the angles in one of the ports of the FOLM where we in which we may have a minimum and maximum of reflectivity the laser cavity. Also that we can select the best performing region in terms of period, amplitude spectrum of the FOLM and by temperature we can shift the wavelength in the FOLM and equalize the two wavelengths required to generate a laser with dual wavelength emission.

In the second part we propose to apply the FOLM to generate a laser with dual wavelength emission. We propose and demonstrate experimentally a laser with dual wavelength and stable, we can make the laser having laser emission at single or dual wavelength by adjusting the temperature in the loop FOLM and we demonstrate how to improve the stability of the laser by adjusting the amplitude using the optical fiber twisters in the FOLM.

In the third part we explain the implementation of the FOLM to generate tunable dual wavelength using a polarizer maintaining fiber Bragg grating (PM-FBG). We propose and demonstrate experimentally a tunable wavelength laser. The tuning range was 8.06-nm; this tuning was achieved by stretching and compressing the PM-FBG. For each tuning was only necessary to adjust the temperature in the FOLM. As a result of this application of the FOLM to generate a dual wavelength laser, we present two simple configurations that can be used for future applications.

ACKNOWLEDGEMENTS

This work is supported by CONACYT grant 151434.

REFERENCES

1. Talaverano L., Abad S., Jarabo S., and Lopez-Amo M. Multiwavelength fiber laser souces with Bragg-grating sensor multiplexing capability. J. Lightwave Technology 2001; 19(4) 553-558.

2. Liu D., Ngo N. Q., Tjin S. C., and Dong X. A dual-wavelength fiber laser sensor system for measurement of temperature and strain. IEEE Photonics Technology Letters 2007; 19(5) 1148-1150.

3. Mao Q., and Lit J. W. Y. Switchable multiwavelength Erbium-doped fiber laser with cascaded fiber grating cavities. IEEE Photonics Technology Letters 2002; 14(5) 612-614.

4. Zhang H., Liu B., Luo J., Sun J., Ma X., Jia C., Wang S. Photonic generation of microwave signal using a dual-wavelength single-longitudinal-mode distributed Bragg reflector fiber laser. Optics Communications 2009; 282(20) 4114-4118.

5. Liu Z., Liu Y., Du J., Yuan S., Dong X. Switchable triple-wavelength Erbium-doped fiber laser using a single fiber Bragg grating in polarization-maintaining fiber. Optics Communications 2007; 279 168-172.

6. Latif A. A., Ahmad H., Awang N. A., Zulkifli M. Z., Pua C. H., Ghani Z. A., Harun S. W. Tunable high power fiber laser using AWG as the tuning element. Laser Physics 2011; 21(4) 712-717.

7. Han. Y., Lee J. H. Switchable dual wavelength Erbium-doped fiber laser at room temperature. Microwave and Optical Technology Letters 2007; 49(6) 1433-1435.

8. Ahmad H., Zulkifli M. Z., Norizan S. F., Latif A. A., Harun S. W. Controllable wavelength channels for multiwavelength Brillouin Bismuth/Erbium based fiber laser. Progress in Electromagnetics Research Letters 2009; 9 9-18.

9. Ahmad H., Sulkifli M. Z., Thambiratnam K., Latif A. A., Harun S. W. Switchable semiconductor optical fiber laser incorporating AWG and broadband FBG with high SMSR. Laser Physics Letters 2009; 6(7) 539-543.

10. Feng S., Xu, O., Lu S., Mao X., Ning T., Jian S. Switchable dual-wavelength Erbium-doped fiber-ring laser based on one polarization maintaining fiber Bragg grating in a Sagnac loop interferometer. Optics and Laser Technology 2009; 41 264-267.

11. Ahmad H., Zulkifli M. Z., Thambiratnam K., Latif A. A., Harun S. W. High power and compact switchable Bismuth based multiwavelength fiber laser. Laser Physics Letters 2009; 6(5) 380-383.

12. Liu D., Ngo N. Q., Chan H. N., Teu C. K., Tjin S. C. A switchable triple –wavelength Erbium-doped fiber laser with a linear laser cavity. Microwave and Optical Technology Letters 2006; 48(4) 632-635.

13. Feng S., Xu O., Lu S., Ning T., Jian S. Switchable multi-wavelength Erbium-doped fiber laser based on cascaded polarization maintaining fiber Bragg gratings in a Sagnac loop interferometer. Optics Communication 2008; 281 6006-6010.

14. El-Tasher A. E., Alcon-Camas M., Babin S. A., Harper P., Ania-Castañon J. D., Turitsyn S. K. Dual-wavelength, ultralong Raman laser with Rayleigth-scattering feedback. Optics Letters 2010; 35(7) 1100-1102.

15. Parvizi R., Harun S. W., Ali N. M., Shahabuddin S., Ahmad H. Photonic crystal fiber-based multiwavelength Brillouin fiber laser with dual-pass amplification configuration. Chinese Optics Letters 2011. 9 021403.

16. Im J. E., Kim B. K., Chung Y. Tunable single and dual-wavelength Erbium-doped fiber laser based on Sagnac filter with a high-birefringence photonic crystal fiber.

17. Liu D., Ngo N. Q., Liu H., Liu D. Microwave generation using an all polarization-maintaining linear cavity dual-wavelength fiber laser with tunable wavelength spacing. Optics Communications 2009; 282 1611-1614

18. Moore P. J., Chaboyer Z. J., Das G. Tunable dual-wavelength fiber laser. Optical Fiber Technology 2009; 15 377-379.

19. Alvarez-Tamayo R. I., Duran-Sanchez M., Pottiez O., Kuzin E. A., Ibarra-Escamilla B., Flores-Rosas A. Theoretical and experimental analysis of tunable Sagnac high-birefringence loop filter for dual-wavelength laser application. Applied Optics 2011; 50(3) 253-260.

20. Ma X., Kai G., and Wu Z. Study of polarization independence for high birefringence fiber Sagnac interferometers. Microwave Optical Technology Letters 2005; 46 183–185.

21. Liu L., Zhao Q., Zhou G., Zhang H., S Chen., Zhao L., Yao Y., Guo P., Dong X. Study on an optical filter constituted by concatenated Hi-Bi fiber loop mirrors. Microwave Optical Technology Letters 2004; 43 23–26.

22. Lim K. S., Pua C. H.,N. Awang A., Harun S.W., Ahmad H. Fiber loop mirror filter with two-stage high birefringence fibers. Progress in Electromagnetics Resarch C 2009; 9 101–108.

23. Li S., Ngo N. Q., Tjin S. C., Binh L. N. Tunable and switchable optical bandpass filters using a single linearly chirped fiber Bragg grating. Optics Communication. 239 (2004) 339-344.

24. Moon D. S., Sun G., Lin A., Liu X., Chung Y. Tunable dual-wavelength fiber laser based on a single fiber Bragg grating in a Sagnac loop interferometer. Optics Communications 2008; 281 2513-2516.

25. Mirza M. A., Stewart G. Theory and design of a simple tunable Sagnac loop filter for multiwavelength fiber lasers. Applied Optics 2008; 47 5242–5252.

26. Sun H. B., Liu X. M., Gong Y. K., Li X. H., Wang R. Broadly tunable Dual-Wavelength Erbium-Doped Ring Fiber Laser Based on a High-birefringence Fiber Loop Mirror. Laser Physics 2010; 20(2) 522-527.

27. Zhou K. J., Ruan Y. F. Fiber ring laser employing an all-polarization-maintaining loop periodic filter. Laser Physics 2012; 20(6) 1449-1452.

28. Mortimore D. B. Fiber loop reflectors. J. Lightwave Technology 1988; 6(7) 1217–1224.

29. Kuzin E.A., Cerecedo-Nuñez H., Korneev N. Alignment of a birefringent fiber Sagnac interferometer by fiber twist. Optics Communications 1999; 160 37-41 (1999)

30. Durán-Sánchez M., Flores-Rosas A., Alvarez-Tamayo R. I., Kuzin E. A., Pottiez O., Bello-Jimenez M., Ibarra- Escamilla B. Fine adjustment of cavity loss by Sagnac loop for a dual wavelength generation. Laser Physics 2010; 20(5) 1270–1273.

31. Alvarez-Tamayo R. I., Durán-Sánchez M., Pottiez O., Kuzin E. A., Ibarra-Escamilla B. Tunable Dual-Wavelength Fiber Laser Based on a Polarization-Maintaining Fiber Bragg Grating and a Hi-Bi Fiber Optical Loop Mirror. Laser Physics 2011; 21(11) 1932-1935.

Electrochemical Scanning Tunneling Microscopy (ECSTM) – From Theory to Future Applications

Ajay Kumar Yagati[1], Junhong Min[3],
and Jeong-Woo Choi[1, 2]

[1]Research Center for Integrated Biotechnology, Sogang University, Seoul, Republic of Korea

[2]Department of Chemical and Biomolecular Engineering, Sogang University, Seoul, Republic of Korea

[3]School of Integrative Engineering, Chung-Ang University, Seoul, Republic of Korea

INTRODUCTION

The development of scanning tunneling microscopy (STM) clearly forms the creation of a new research tool by innovative implementation of scientific and technological knowledge, thereby advancing further in the fundamental science and technology [1,2]. The quantum-mechanical phenomenon of electron tunneling had been known for a long time, but the use of this phenomenon for the imaging of a conductive surface at atomic level was realized only in 1982 when the first STM as built by Binnig et al [3]. STM has a resolution of a few Ångstrom in lateral directions and less than one Ångstrom in the direction perpendicular to the surface [4]. It consists of a scanning tip which images the surface by means of a tunnel current. Hence, the sample needs to be conductive [5]. At present, STM is a powerful tool for analyzing metallic and semiconductor surface. The real-space visualization of surface at atomic scale is one of the most important features [6]. The spatial variation of the tunneling current or the spatial variation of the tip height is converted in to the real space image. The tunneling current decreases exponentially with the increase in tip-sample distance. Thus, at any given location of tip over the sample surface, the electron transfer involves only one atom or few atoms at the tip apex and on the surface closest to them. This makes it possible to visualize the structures with sub-angstrom resolution and to detect atomic scale defects that are not possible with other spectroscopic techniques [7,8]. The STM not only provides the three-dimensional information about the topography of the sample, but it also gives the information about the spectroscopic properties and local variations of work functions. Further, as a nanofabrication tool, STM can be used for atom manipulation, local deposition and imaging of the molecules [9]. Moreover, STM can be used to operate in air, ultra high vacuum (UHV) and in liquid solutions for electrochemistry applications which involves the immersion of the STM probe into the liquid media and the corresponding electrochemical control (ECSTM) for in situmonitoring of redox processes on the sample electrode [10,11]. Having these advantages STM has become most widely accepted analytical measurement system in the current research works.

FUNDAMENTALS OF SCANNING TUNNELING MICROSCOPY (STM)

Origin and Operation Mode:

The scanning tunnelling microscope was developed by Binnig, Rohrer, Gerber and Weibel [12]. Since STM can be used for imaging on at atomic scale level, this belongs to the most powerful experimental techniques of surface science. In STM, a sharp metallic tip is placed very close to the surface and a small bias voltage is applied between the tip and the sample. As a result, a current of electrons (IT),flows between the electrodes through the vacuum gap. This process is a quantum mechanical phenomenon and is called as "tunnelling" effect [13] shown in Fig. 1. The electrons "tunnel" through this electrically insulating layer, giving rise to a measurable current which displays an exponential dependence on the distance between the two conducting electrodes [14]. The tunneling current flowing between the STM tip and the sample surface through the insulating gap (s) under an applied V_{bias} which can be explained in a simple analytical tunneling expression assuming a 3-dimensional (3D) metal-insulator-probe junction into a one-dimensional metal-insulator-metal contact. The derived equation that relates with the applied V_{bias} with I_T and s yields [15]:

$$I_T = I_o \frac{V_{bias}}{s} e^{-k\sqrt{\phi_b}S}$$

(1)

It is the exponential dependence of the tunneling current on the distance between the two conductors (STM tip and the underlying surface) that provides the sensitivity of the measured current that can be interpreted as the surface structure. As the equation shows, other factors influence the current as well, such as the electron band structure of the two conductors (φ).

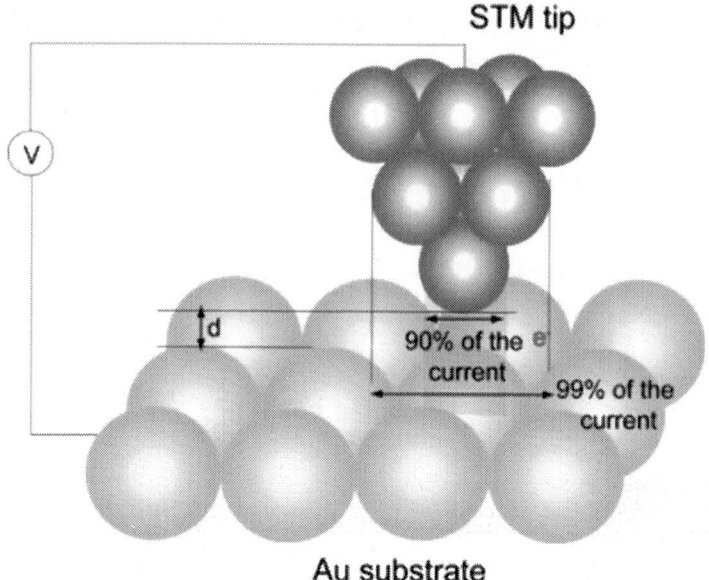

Figure 1: Schematic diagram of scanning tunneling microscopy depicts the tip sample interaction and the tunneling current between the tip and the sample surface.

Tunnelling Methods

General Modes of Operation

Basically there are two modes in which STM can be operated. The first one being the constant current and the other one is termed as constant height. In the constant current mode, the tip is scanned over the surface at constant tunnel current, and the vertical tip position will be continuously changed to keep the tunnel current as a constant. In ideal conditions, at a homogenous surface constant current refers to constant interval between sample surface and the tip. In this mode, the height control mechanism will adjust the tip to move vertically up and down to keep the tunnel current to a constant value by the feedback voltage [16]. In the constant height mode which most effective for the investigation of atomically smooth surfaces (Z remains const). In this mode of operation, the tip moves above the surface at a distance of several Å, and the changes in the tunneling current are recorded as STM image [17]. Scanning may be done either with the feedback system switched off (here no topographic imaging is recorded), or at a speed exceeding the feedback reaction speed (only smooth changes of the surface topography are recorded). This method employs very high scan rates and fast STM images acquisition, allows observing the changes that occur on a surface in a real time.

The Electrochemical STM (ECSTM) Model

The electrochemical scanning tunneling microscopy (ECSTM) is an extended technique performed along with basic STM measurements for the study of electrode-electrolyte interfaces. Therefore, the same elements as in a standard STM experimental set up can be observed in the ECSTM set-up. The tunneling current flowing between a metallic tip and a conductive sample will be again used to obtain topographical information as well as the electronic structure of a determined electrode surface immersed in the corresponding electrolyte [18,19]. Even though, the fundamentals of both techniques are essentially the same, but two different elements were introduced in a conventional STM set up in order to control as an ECSTM: 1) a three electrode electrochemical cell in which the substrate used as a working electrode, a reference and a counter electrode completed the electrochemical cell. This configuration resembles as a normal three electrode configuration coupled to a potentiostat in which the potential of the working electrode can be controlled with respect to a high-impedance reference electrode, while the current is allowed to flow between the working and the counter electrode. 2) Development of suitable ECSTM probes. In this case, the tip is not an active electrode in the cell, but used only to image the surface morphology, even if it had to be under potential control in order to apply a voltage drop to drive the tunnelling current. The implementation of a liquid STM represented a great breakthrough for the in situ study of surface electrodes [20]. An improvement in the STM has been obtained with the introduction of the bipotentiostat approach, in which both tip and sample potentials are independently controlled with respect to a reference electrode in solution.

ELECTROCHEMISTRY IN SCANNING PROBE MICROSCOPY: BASIC CONCEPTS AND APPLICATIONS

The electrochemical scanning tunneling microscope was the first tool for the investigation of solid-liquid interfaces that allowed in situ real space imaging of the underlying electrode surfaces at atomic level. Therefore ECSTM gained much importance and emerged as a prominent tool for the determination of the local surface structure as well as the dynamics of reactions/ processes that takes place at surfaces in an electrolytic environment. Although the fundamentals of both techniques are then essentially the same, but two different elements must be introduced in a conventional STM set up in order to operate as an ECSTM: an electrochemical cell consisting of two working electrodes with bipotentiostat approach and suitable ECSTM probes [21].

Preparation of Reliable Probes, Tunneling in Liquid Environments at a Bipotentiostat Configuration

The general electrochemical experiment which is composed with standard potentiostat, the potential of the working electrode is controlled with respect to a reference electrode by the flow of current through a counter electrode. However, in ECSTM with in-situ electrochemical measurements, the same potentiostatic approach is able to control both the potential of the substrate and the potential of the tip that is present in solution with respect to a reference electrode in order to control electrochemical reactions taking place at its surface, which is practically not possible [22,23]. Generally in STM, it is expected that the current enters into the measuring system due to the charge transfer process is mainly due to tunneling current between the tip and the substrate. If the potential of one of these two electrodes is not controlled in the electrochemical cell, then electrochemical charge transfer mechanisms might become significant and contribute to the measured currents which leads to a strong source of noise in the STM control circuit [24,25]. Further, the absence of control on the electrochemical potential of one electrode in an electrochemical cell also gives rise to phase variation in the surface composition of the electrode. In this aspect, with bipotentiostatic approach an independent control on the potential of both the tip and the substrate with respect to a reference electrode in solution is established for the development of the in-situ STM.

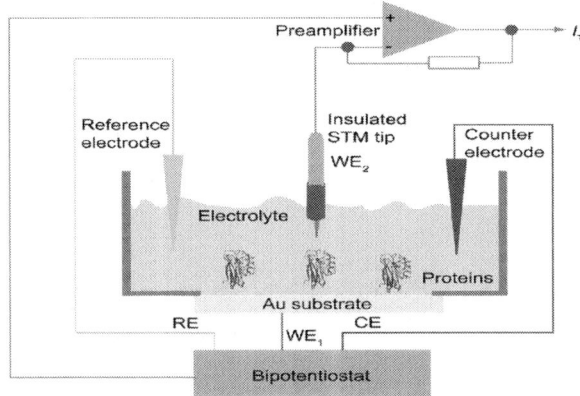

Figure 2: EC-STM configuration, the bipotentiostat controls potential of tip and sample with respect to reference electrode.

The main purpose of the potentiostat is to control the potential of the electrode in an electrochemical cell from various impedances connecting to these electrodes. The controller maintains the potential of the reference with respect to the working electrode in such a way that the potential is exactly opposite to the controlled potential, which is free from fluctuations of the impedances [26]. Hence, the bipotentiostat controls the potential of two electrodes with respect to a reference electrode. The ECSTM setup is depicted in Fig. 2. Generally, the tip is virtually grounded and the tunneling current is measured by a high-gain current follower or it is fed to the STM control unit through a preamplifier. In all configurations, the tip and substrate potentials are controlled with respect to a current less reference electrode. Usually, the counter electrode in EC-STM setup is obtained by an Au or Pt wire which has enough stability [27]. Also, a good reference electrode is obtained when a metal is used in electrochemical equilibrium with the corresponding metal cation in solution. In the past decade,

STM has been applied to study the solid-liquid interfaces, in which the STM probe is immersed into the liquid media and controlled by the electrochemical module of STM for in-situ monitoring of redox process on the sample electrode. ECSTM has advantages over classical or conventional STM operated either in air or in a vacuum. It allows precise and independent control of both tip potential and sample electrode potential through the use of a bipotentiostat and a quasi-reference electrode [28]. Generally, STM allows to perform electronic spectroscopy by recording the tunneling current (I_t) with the applied bias potential (V_b) known as scanning tunneling spectroscopy (STS) and the same can be applied in electrochemical environment, leading to electrochemical scanning tunneling microscopy (ECSTS) technique [29]. Even though ECSTS is a sophisticated and powerful tool it is not widely used because of complexity in tip preparation methods.

Generally in ECSTS, the current flowing through the tip has several components; a) STM tunneling current, b) faradaic current by the electrochemical reaction at the tip/electrolyte interface and c) charge-discharge process of the electrochemical double layer at the tip-electrolyte interface. So, if the faradaic currents and the charge-discharge processes are larger than the set-point tunneling current, the STM measurement will no longer be possible. Hence there should be an alternative to eliminate these two electrochemical contributions to the measured tip current. The most effective alternative is, taking the advantage of the fact that faradaic/capacitive currents are directly proportional to the exposed area of the electrode, coating the tip except for the very apex so that tunneling current can flow while faradaic and capacitive currents can be minimized [30,31]. Generally in STM, the tunneling current is usually set to between 1~10 nA. For accurate measurements, coated tips must yield faradaic and capacitive currents ≤ 0.1 nA at the end of the STS curve when ramping at the highest tip potential scan rate (up to 10 V/s) needed to minimize drift, whereas the STS curve is recorded without feedback. Various methods and materials have been proposed for coating purposes, such as apiezon wax, melt glass, copolymers, and a combination of glass and polymer, and in all these coating procedures, the faradaic and capacitative contributions are too large to be subtracted from the measured tip current, and thus, they do not allow STS spectra to be under electrochemical control [32].

As mentioned above, the electrochemical charge associated to the charging-discharging process of an electrode in contact with a specific electrolyte depends directly on the exposed area of the electrode itself. For this purpose, the ECSTM tips must be insulated from the electrolyte, in the way that just the very end tip apex remains in contact with the electrolyte. With this isolation process, the electrochemical current measured through the STM must be better than 10 % of the tunneling set point current, that is, typically $\leq$ 0.1 nA. Electrophoretic paints are increasingly used for coating Pt-It tips as they are chemically and electrochemically inert insulators [33]. However, among many protection methods available, apiezon wax is still used for ECSTM tip isolation, despite its disadvantages.

CHARACTERIZATIONS WITH ECSTM

Understanding Electrochemical Processes Such as Corrosion, Deposition, and Adsorption

Applications of ECSTM in the field of corrosion have mainly focused on understanding the mechanism of corrosion initiation and the process of inhibition, including pitting initiation, surface dissolution, passive film formation, and the effect of inhibitors. ECSTM has been utilized to study a variety of materials, including Cu, Ni, and Fe, in many different corrosion environments [34]. In general, the termcorrosion stands for material deterioration or surface damage in a liquid environment. In the case of metals, it is basically a chemical/electrochemical process that suggests an oxidation of a metal which transfer electrons to the electrolytic environment and undergoes a change of valence from zero to a positive value [35]. Normally, this initial process leads to a number of parallel electrochemical methods that result in material dissolution and/or eventual formation of secondary corrosion products. Basically the metal corrosion processes can be classified in two different forms: one is general corrosion [36] that symbolizes those corrosion processes involving the entire surface area of the material, and the second one localized corrosion [37] which mentions to a number of corrosion processes that are triggered at specifics sites on the material surface. The presence of a surface passive film prevents the metallic substrate from further oxidation, i.e., the surface becomes passivated. However, the passive state of a metal under certain conditions is susceptible to localized instabilities that prompt the corrosion of the metal electrode through the local dissolution of the passive layer. This process is called as pitting corrosion [38].

Pitting corrosion takes place at passivated metal surfaces and leads to a creation and growth of an activepit [39]. It is quite common that these pits can easily develop at defective sites on the passive film surface, on sharp edges or sites where the oxide film is thinner. The essence of the pitting process relays in its anodic nature (metal dissolution) compared with a passivated cathodic region surrounding it, thus allowing the continuous flow of electrons to take place [40]. The pit formation is depicted in Fig. 3 where peaks indicate the pitting corrosion. Most metal passive layers present a semiconducting or insulating electronic behavior and, therefore, it is expected to understand the initiation of the pitting corrosion process, this requires to embed semiconductor phenomena into an actual classical corrosion mechanisms.

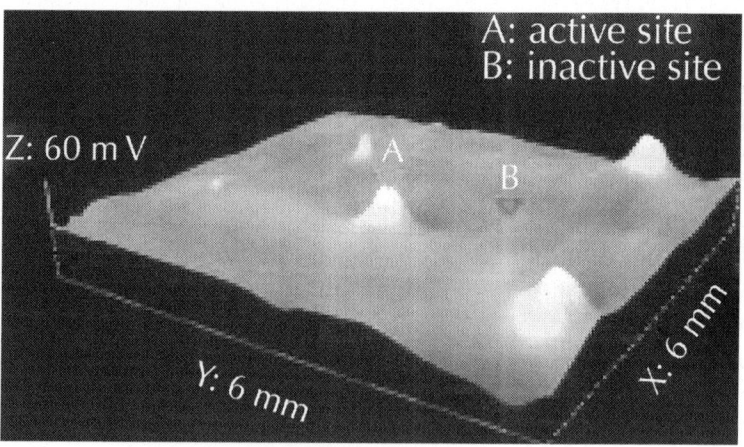

Figure 3: The typical 3D potential distribution image for the stainless steel in 10% (wt%) $FeCl_3$ solution. Figure reproduced with permission from: ref. 43, © 2012 Elsevier.

It is known that, charge injection always occurs at high electric fields in particularly under dc conditions. The presence of space charge in solid dielectrics will result in distortion of electric field distribution. Dielectric breakdown can began in the region where electric field enhancement takes place if the electric field exceeds the "threshold strength" of the material. That is, spontaneous generation of extra charge carriers in an electrically insulated region either by tunneling (Zener breakdown) or by collision (avalanche breakdown) [41]. Both these mechanisms occur when a sufficiently high electric field is applied to the oxide layer thus generating high currents passing through the interface.

The location of a carrier at the semiconductor solid surface is generally accounted as the lattice bond weakening that ultimately lead to the bond breaking and the corresponding electrochemical electrode corrosion (dissolution). This process may continue in different routes that comprise either the movement of electrons and holes to charged states [42]. The rate of these charge carriers (e^- and h^+) arriving to the particular charges state will then govern the dissolution rate and, consequently, the material corrosion. ECSTM can be used to analyze the pitting corrosion by measuring potential distribution image for the stainless steel in 10% (wt.%) $FeCl_3$ solution after 30 min of immersion [43]. Further, ECSTM is able to map the in-situ pitting process with micro and nano-spatial resolution and can locate the positions also the local activity of pitting corrosion in an early stage.

ECSTM has been a valuable tool for understanding electrochemical processes such as corrosion, deposition and adsorption. For example, a method to estimate the pitting corrosion of naked and self-assembled monolayer (SAM) of n-alkane thiol modified Au (111) surfaces in CN^- solutions was examined [44]. It is estimated that SAMs reduced the rate of corrosion, but fragile and contain too many defects. To estimate this corrosion, the applied potential was slowly changed from negative to positive values and it is observed that at small positive potential values leads the initial stage of corrosion, at slightly higher positive potentials pits and step edges occurs and even higher potentials the etching starts and surface becomes rough [45]. Overall, the SAM modified surface is more resistance to corrosion for longer periods of time and at more positive (etching) potentials than the naked Au surface. Hence, 1) the potential imaging technique is able to locate the positions and map local activity of the pit initiation. Hence ECSTM study becomes an alternative method for analyzing the pitting initiation at an early stage; 2) combined with potential imaging, in situ ECSTM can be useful for imaging the surface topographies that associated with dynamic process of local breakdown of passive layers and micropitting initiation; 3) a threshold potential can be defined as a critical potential to characterized the local breakdown of passive layers; 4) the pitting initiation is strongly depends on the surface conditions of passive film, concentration of the ions and pH in the solution.

Molecular Resolution Imaging of Redox Species in Solution with ESTM

In biological molecules, electron transfer (ET) in solid surface or liquid solution can occur between donor (D) and acceptor (A) separated by a long distance. To describe this process quantitatively, sophisticated models are required. Additionally, scanning tunneling and atomic force microscopy (STM and AFM, respectively) have opened an exciting new perspective for molecular imaging. STM imaging at the solid/air interface to molecular and occasionally sub molecular resolution also has been extended to biological macromolecules including DNA and a number of redox and non-redox proteins [46,47].In-situ STM offers, on the other hand, new electrochemical spectroscopic probes in addition to current-bias voltage relations, particularly the relation between the tunnel current and the overvoltage of both the tip and substrate electrodes relative to a common reference electrode. Particularly, in-situ scanning tunneling microscopy (STM) of redox molecules, in aqueous solution, shows interesting analogies and differences compared with interfacial electrochemical electron transfer and also in homogeneous solution. With ECSTM, high resolution imaging and spectroscopy of adsorbed molecules can be achieved. It is understood that ECSTM combines electrochemical control and STM high-resolution profiles such as molecular imaging and scanning tunneling spectroscopy (STS) [48]. The possibility of using ECSTM to observe single-molecule charge transport was proposed in the early 1990s. The first ECSTM experiments was performed on iron porphyrin molecules adsorbed on highly ordered pyrolytic graphite (HOPG) surface, and the redox-tuned resonant tunneling effect was directly visualized by STM imaging [49]. Since then, it has become a powerful tool to study the interfacial electron transfer and molecular conductance of electro active species at single-molecule level in an electrochemical environment. For conductance imaging, with in situ ECSTM mapping was performed on redox molecules such as azurin (a redox protein) seen in reference [50]. To observe the electron transfer properties and its imaging, Azurin (Az) is a blue single-copper protein adsorbed on gold surface which functions as an electron carrier physiologically associated with oxidative stress responses in bacteria (e.g., Pseudomonas aeruginosa) and is a long-standing model for exploring electron tunneling through protein molecules was studied. The molecular assembly for the effective coupling of Azurin with gold surface can be examined with STM imaging or by examining the electrochemical property of the adsorbed Az on Au surface. Moreover, the combination of electrochemical and STM measurements thus provide a way to quantify one of the fundamental and long-lasting questions in adsorption chemistry of redox proteins: Such as mainly 1) what percentage of the protein molecules retains their biological activity in the immobilized state? 2) In addition to fast ET, stability is another key factor determining reproducibility and operation in applications of the system to molecular electronics. ECSTM enables to observe single-molecule current-voltage relations by tuning the over potential across the equilibrium redox potential. The energy state of both the substrate and the tip in ECSTM is under control by electrochemical potentials relative to a common reference electrode in an aqueous buffer environment [51,52]. This STM configuration is particularly suitable for in situ mapping of electronic properties of redox proteins during their biological action (e.g., ET or electrocatalysis), because the aqueous phase is essential for almost all biological processes in nature. STM imaging was performed to observe single molecules, for which high-resolution images of Az was obtained by keeping a constant bias voltage between the substrate and the tip, with the substrate potential set at the equilibrium redox potential (zero over potential) of azurin. Imaging was performed toward either positive or negative overpotentials by adjusting the substrate and tip potentials in parallel (i.e., at constant bias voltage) and finally was returned to the equilibrium potential [53]. The adsorbed azurin monolayer is sufficiently robust and can withstand repeated ECSTM imaging without loss of its activity. Hence, a series of STM images was obtained at various overpotentials. Fig. 4 shows typical images in which three molecules were targeted. Focus is on the central molecule; two molecules in the upper left region serve as a positioning reference. The single-molecule contrast is clearly tuned

molecular energy levels can be mostly located either above or below the tip Femi level, resulting in no significant resonant tunneling tuned by the substrate potential.

The tunneling current (I_t) dependence on the effective overpotential ($e\xi\eta$) and the bias voltage (eV_{bias}) can be analysed in steady-state electron transfer form in the following combination [50].

$$I_t = 2en\frac{k^{o/r}k^{r/o}}{k^{o/r} + k^{r/o}}$$

(2)

where $k^{o/r}$ and $k^{r/o}$ are the rate constants, for electron transfer between the tip and protein and between the protein and substrate respectively. n is the number of electrons transmitted in a single electron transfer event and e the electronic charge.

$$k^{o/r} = \kappa_t\rho_t\frac{\omega_{eff}2k_BT}{\alpha_t}\exp\left(-\frac{\left(\lambda - e\xi\eta - e\gamma V_{bias}\right)^2}{4\lambda k_BT}\right)$$

(3)

$$k^{r/o} = \kappa_s\rho_s\frac{\omega_{eff}2k_BT}{\alpha_s}\exp\left(-\frac{\left(\lambda - eV_{bias} + e\xi\eta + e\gamma V_{bias}\right)^2}{4\lambda k_BT}\right)$$

(4)

where κt and κs are electronic transmission coefficients for electron transfer between the tip and the protein, and between the protein and the substrate, respectively; ρt and ρs are the electronic level densities of the tip and the substrate; ω_{eff} is the effective nuclear vibrational frequency, αt and αs are the transfer coefficients for electron transfer between the tip and the protein and between the protein and the substrate, respectively; λ is the reorganization free energy; ξ is the fraction of the substrate-solution potential drop, η is the overpotential;

A mediator is added (concentration in mM) to the supporting electrolyte that is converted at the UME under diffusion controlled conditions which produces a steady-state current, I_T,∞, in the bulk phase of the solution. If the UME is brought close to an inert and insulating surface, it blocks the diffusion of the mediator to the UME and the UME current I_T decreases below I_T,∞ (called negative feedback) [68]. If the UME is brought above a conductive surface or a catalytically active surface, I_T increases above the value found for an inert and insulating surface. The magnitude of this increase depends on the local reactivity of the sample. A diffusion-controlled reaction at the sample and the tip constitute an important limiting and is called as positive feedback [69].

- Advanced tip positioning: SECM tip usually needs to be positioned close to an interface with high precision. Accurate positioning is achieved by attaching the tip to piezoelectric translators. However, this still leaves the problem of determining the exact separation of the tip electrode and the surface commonly known as 'distance of closest approach' of the electrode with the surface [70]. One can use the amperometric response of the tip electrode in some instances for many systems it might be difficult to add a redox-active species to the solution, without affecting the process or the viability of the sample. Also there are challenges that include in low analytes concentrations or background processes in biological media which means it is hard to measure the distance accurately from the amperometric response [71]. Hence, much effort has been directed towards the development of alternate procedures for tip positioning and distance determination.

Shear force modulation is one method to achieve the control of tip-sample separation by shaking the electrode through a small oscillation in the x-y plane. As the electrode is brought close to a surface, the oscillation is damped, to a degree which depends on the tip-substrate separation. Images are usually acquired at constant damping amplitude, which resembles to a constant distance between the tip and substrate; thus, the tip follows the surface contours [72]. Further, Tip position modulation SECM refers to an operation where an amperometric tip is oscillated in a sinusoidal motion perpendicular to the surface. The resulting current varies with the frequency of the driving oscillation. The amplitude and phase of the oscillating current enable one to deconvolute the activity and topography of the surface [73]. The phase of the current is the same as the phase of the tip-surface separation when the probe is oscillated above an inert surface, whereas they are entirely out of phase above a conducting surface (in positive feedback mode).

- SECM with ECSTM: SECM coupled ECSTM has the dual benefits of nanometer scale resolution imaging along with the ability of the electrochemical measurements. For example see [74]. In order to perform measurements with SECM at nanometer scale many number of technical hurdles has to overcome. Some of them are 1) preparation of suitable nanoscale electrodes with insulator coatings 2) scanning and positioning the probe above the sample surface with a sample distance of 10 nm which is much larger than the tunneling distance but smaller than some electrode radii of the probe, while avoiding mechanical contact between sample and probe. In order to observe electrochemical reactivity at individual nanometer-sized features, such features have to be prepared on the sample surface with such a large distance that SECM can determine the signals of individual features. But it is very difficult to avoid the possibility of mechanical contact between the probe and the surface structures. Hence a method is adopted in which the probe is used in ECSTM mode over the protruding regions of the sample surface and retracted from there [75]. A novel instrumentation has been developed for the imaging with ECSTM and SECM with Pt/Ir wire coated with paint is used as the tip. The tip is kept at a working distance of 20 nm which is much larger than the tunneling distance but enough for feedback imaging for SECM. Probes for the operation of ECSTM and SECM is the most important factor so selecting a desired probe is crucial for better imaging and electrochemical analysis. Tips can be distinguished by the geometric shape of the active electrode area and the insulating sheath for different applications.

Disk-shaped microelectrodes which are the preferred shape for quantitative SECM experiments in the micrometer range, which consists of a Pt wire, sealed in glass or quartz with a laser-heated capillary puller (Fig. 7a) [76]. Because of the large insulating sheath these electrodes are unlikely to function as ECSTM probes. Ring-disk electrodes have been approximated as modified cantilevers for combined scanning force microscopy (SFM)/SECM experiments (Fig. 7b) [77]. The probes have been produced by modification of SFM cantilevers and shaping individual cantilevers by fast ion bombardment. The outer electrode was square-shaped with a side length of 1.5 mm.

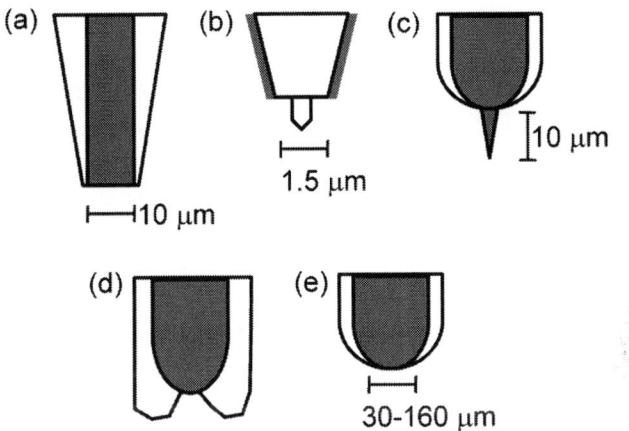

Figure 7: Schematics of the shapes of different microelectrodes used in scanning probe microscopy. Figure reproduced with permission from: ref. 40, © 2010 Elsevier.

Electrodes shown in Fig. 7c are generally used for high-resolution ECSTM experiments. The pointed tip allows atomic resolution in tunneling experiments. However, the active electrode area is decreased by an insulating coating which leaves the pointed area of about 10 mm length open. This is sufficient because the potentials for the imaging experiments are selected such that no Faradic reactions proceed at the tip potential. The tunneling current can be as high as 1 nA so that other currents do not interfere significantly with the experiments. Such electrodes are, however, not useful for SECM feedback experiments. Mediators can access the UME by diffusing parallel to the sample so that no diffusional blocking occurs above passivated samples [78]. The positive feedback would be insignificant compared with the large background current from mediator conversion at such electrodes. Electrodes with a slightly recessed active electrode area (Fig. 7d) were used for amperometric single molecule detection. No images were recorded with such probes. For the combined ECSTM/SECM operation the electrodes such as in Fig. 7e are required. Although they are not as pointed as the one in Fig. 7c, they should allow ECSTM experiments of flat samples although not with atomic resolution. Apart from that, the shallow cone would still provide a blocking of the mediator diffusion above passivated samples [79,80].

Operating principle:

The combined ECSTM/SECM operation is based on the sequential acquisition of ECSTM and SECM [74] data shown in Fig. 8. Once the probe is brought in tunneling contact, an ECSTM image is recorded in the constant current mode. It provides topographic data of the sample. After completion of the ECSTM scan, the electronic feedback loop is switched off and the probe is retracted 20 nm from the working point of the ECSTM scan. This distance is much larger than the tunneling distance. At the same time the potentials at the probe are switched for a desired range to obtain the redox properties of the adsorbed molecule on the surface. After completing the SECM scan, the probe is brought back into tunneling contact with the sample and a step perpendicular to the high-frequency scan axis is performed. From there the sequence is repeated until a full image frame is recorded.

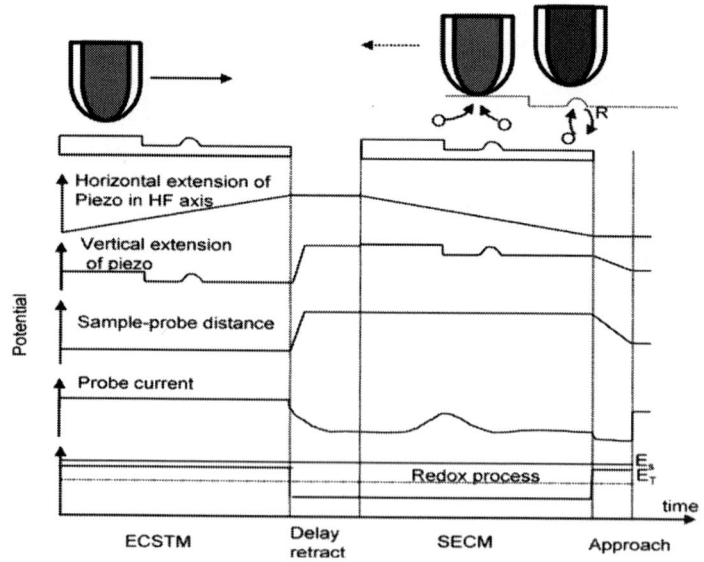

Figure 8: Schematic diagram for the combined operation of ECSTM-SECM performance. Figure reproduced with permission from: ref. 40, © 2010 Elsevier.

Limitations of ECSTM

ECSTM has several limitations and drawbacks, generally these limitations resides in the difficulty to perform bias-dependent measurements such as when potential of the substrate is fixed during electrochemical process then the tip is adjusted to maximize the faradic current and to optimize STM measurements, with the tunnel voltage is no longer adjusted over long range. Hence in STS measurements, voltage dependent imaging is not possible to measure in ECSTM mode [81]. Limitations are also arises from the possible interference of the tip with the electrochemical process at the working electrode. The close proximity of the tip causes shielding effects for reaction at the sample-solution interface.

Further, corrosion studies by ECSTM are also facing problems as high rate of mass transport with respect to time for acquisition of images thus leads to low resolution images. Apart from these limitations, ECSTM is an effective tool which can provide valuable information related to molecular structures of the adsorbed molecules at nanoscale and also provide information relating to corrosion process like adsorption, dissolution and localized corrosion of metallic materials.

Instability and Drift

Repeated electrochemical experiments on the all the coated tips leads two typical instability behaviors. A progressive increase of the maximum current is observed on aging, which is associated with dissolution of the coated layer that leaves more and more metal surface uncoated in time. Secondly, all the current levels at each potential drift in time toward higher values, which means that the effective electrical resistance at the tip due to coating decreases. Probably, the solution was leaking into the interface between metal and coating [82].

Mechanical drift is a critical parameter in high resolution imaging caused by the differential thermal expansion of individual instrument components. Sources of drift in the X-Y plane consist of these thermal effects and hysteresis in the piezo-scanner. The Z-direction presents a more complex situation due to the contributions of temperature, sample tilt in the X-Y plane, and hysteresis. Further, in the cases of in-situ and electrochemical experiments where drift is typically higher

[83], the effects of time and evaporation of the solvent must be carefully balanced. The influence of drift in the X-Y plane detrimentally affects two main aspects of the imaging process. First, the ability to image a given region or feature over time is limited. Second, high-resolution imaging of atomic and molecular lattices demonstrates curvature in the observed periodic structure. The analysis of drift in the X-Y plane is carried out by the collection of sequential images [84].

The major effects of drift in the Z-direction are most disruptive during surface spectroscopic measurements. In the STM application, tunnel current versus distance curves (I-S) collected at a constant bias voltage provide a direct measurement of the effective local barrier height involved in the tunneling process. This value is generally unknown and directly influences the more common bias-dependent (I-V) spectroscopy. It is therefore desirable to minimize differential drift in the Z-direction between the tip and sample. Measurement of drift in the Z-direction was quantified through long-term acquisition of the Z-signal in the tunneling condition.

Poorly Resolved Images

High-resolution imaging has been the primary feature that attracted the researcher's attention to scanning probe microscopy yet there are still a number of outstanding questions regarding this function of scanning tunneling microscopes or coupled with other microscopic techniques. Some types of proteins may adhere to the tip. This will reduce resolution giving "fuzzy" images. If tip contamination is suspected to be a problem, it will be necessary to protect the tip against such contamination [85]. The ECSTM images become unclear and noisy at potentials more negative potentials due to the perturbation of the tip caused by severe reactions. A great quantity of spots will be observed on all over the surface. In addition, it is know that on the surface of graphite substrate the alkane molecules are assembled in flat-lying lamellar structures. The alkane adsorbates on graphite were analyzed by means of a droplet of saturated alkane solution that is deposited on graphite surface and the metallic tip penetrates this droplet and a molecular adsorbate at the liquid-solid interface until it detects a tunneling current. At these conditions the tip is scanning over the ordered molecular layer in immediate vicinity of the substrate. A specific feature of the STM imaging at the liquid-solid interface is that the probe is surrounded by the alkane saturated solution [86]. Any instability of the imaging and the use of low tunneling gap resistance cause a mechanical damage of the alkane order, and the probe might record the image of the underlying graphite. If the gap is increased again the alkane order is restored. It is difficult to get STM images of "dry" alkane layers on graphite because an occasional damage of the layer is not repairable. This progress relies on instrumental improvements (better signal-to-noise characteristics, low thermal drift, improved detection and control of the tip-sample forces, etc.) and the use of sharp probes [87].

The other issue is related to the better understanding of the nature of atomic-scale resolution in STM. In some cases, imaging provides only the lattice resolution in the contrast with true atomic resolution where a detection of such defects is expected. The imaging of the periodical lattices with the defects can be demonstrated with the results of the computer simulation which revealed that visualization of the defects does not necessarily mean that the surrounding molecular order is correctly reproduced in the images. These findings emphasize a need of a thorough interaction between the experiment and theory in the analysis of the atomic scale data.

FUTURE TRENDS AND APPLICATIONS

The invention of the scanning tunneling microscope had a revolutionary influence on the development of material characterization at nanometer scale. The future of STM with electrochemical control depends on the development of appropriate probes for implementing the electrochemically controlled current sensing atomic force microscopy (ECAFM) [88]. In comparison of ECAFM and ECSTM, it can be said that both methods can measure the electronic properties of single molecules but ECAFM allows the control of force applied by the tip on the molecules adsorbed on the surface, thus allowing measure the mechanical properties of the single molecules. In comparison with STM, Atomic force microscopy (AFM) doesn't require conductive samples and tips. AFM has the advantages of measuring the local forces between the tip and the sample surface, including van der Waals, Born repulsion, electrostatic and magnetic forces, friction and adhesion [89]. In electrochemical applications, ECAFM is often preferred because it is easier to set-up and the obtained topographic information is independent of the conductivity of corrosion materials. Moreover, ECAFM is most often used at sub-micron level, i.e., at a lower level of spatial resolution that does not require preparation of atomically smooth surfaces as for ECSTM studies. Further, ECAFM can also be combined with a variety of other techniques to analyze corrosion and optimize corrosion protection properties [90]. However in comparison with ECSTM, even though ECAFM doesn't require atomically smooth surfaces like ECSTM and also in ECAFM the cantilever-tip assembly doesn't constitute a fourth electrode through which the electrochemical current that would flow, but there is a disadvantage that in the resonant contact mode, the fluid medium tends to damp the normal resonance frequency of the cantilever which is complicated. Further, the noncontact mode is impractical because the van der waals forces are even smaller making it as a big drawback for biological applications. Such instrument would enable better molecular investigation of electron transfer properties thereby maintaining the molecular conformation. To achieve this spectroscopy, specially designed probes such as insulated conductive probes along with bipotentiostat setup essential. The capability of recording atomic scale features with STM has increased the scope of research to develop other types of STM which provide information about the topography and mechanical, magnetic, electrical properties of the surfaces [91]. The development of these new applications is the design of specific probes, having improved spatial resolution, sensing the desired sample properties and operating in different environments. Developing a highly flexible, compact microscopic system can be easily interfaced with commercial control electronics and integrated with inverted optical microscopy. Simple design architecture enables simple exchange of the probe tip, sample, and scanner while enhancing system resonances and resistance to thermal drift. In recent years, the demand of the applications of ECSTM has greatly increased in numerous fields [92]. ECSTM in combination with SECM has the ability to perform local reactivity imaging simultaneously with ECSTM imaging as well as to induce local electrochemical surface modification in the same setup opens up new perspectives for the investigation of heterogeneous reactions in electrocatalysis at metal clusters and in corrosion processes in a new size regime. Hence with the help of STM, complemented by other characterization techniques, it is reasonable to believe that new advance in building specific and functional surfaces can be achieved in the future.

REFERENCES

1. Eigler D.M, Schweizer E.K., Positioning single atoms with a scanning tunnelling microscope. Nature 1990; 344, 524-526.

2. Stroscio J.A, Eigler D.M., Atomic and Molecular Manipulation with the Scanning Tunneling Microscope. Science 1991; 254, 1319-1326.

3. Binnig G, Rohrer H. Scanning tunneling microscopy. Surface science 1983; 126(1-3) 236-127.

4. Stolyarova E, Rim K.T, Ryu S, Maultzsch J, Kim P, Brus L.E, Heinz T.F, Hybertsen M.S, Flynn G.W, High-resolution scanning tunneling microscopy imaging of mesoscopic graphene sheets on an insulating surface. PNAS 2007; 104, 9209-9212.

5. Heim, M, Eschrich, R, Hillebrand, A, Knapp, H.F, Guckenberger, R, Cevc, G. Scanning tunneling microscopy based on the conductivity of surface adsorbed water. Charge transfer between tip and sample via electrochemistry in a water meniscus or via tunneling? Journal of Vacuum Science & Technology B: Microelectronics and Nanometer Structures, 1996; 14, 1498-1502.

6. Cui, X. D, Primak A, Zarate X, Tomfohr J, Sankey O.F, Moore A.L, Gust D, Harris G, Lindsay S.M, Reproducible Measurement of Single-Molecule Conductivity. Science 2001; 294, 571-574.

7. Hallmark, V.M, Chiang s, Rabolt, J.F, Swalen, J.D, Wilson R.J. Observation of Atomic Corrugation on Au(111) by Scanning Tunneling Microscopy. Phys. Rev. Lett. 1987; 59, 2879-2882

8. Bonnell D.A. Scanning tunneling microscopy. Encyclopedia of Materials: Science and Technology 2001, 8269-8281.

9. Garcia R, Yuqiu J, Schabtach E, Bustamante C. Deposition and imaging of metal-coated biomolecules with the STM. Ultramicroscopy 1992; 42-44, 1250-1254.

77. Kranz C, Friedbacher G, mizaikoff B. Integrating an ultramicroelectrode in an afm cantilever: combined technology for enhanced information. Anal. Chem., 2001; 73 (11) 2491-2500.

78. Neufeld, A.K., O'Mullane, A.P. Effect of the mediator in feedback mode-based SECM interrogation of indium tin-oxide and boron-doped diamond electrodes. J Solid State Electrochem. 2006; 10, 808-816.

79. Szamocki R, Velichko A, Holzapfel C, mucklich F, Ravaine S, Garrigue P, Sojic N, Hempelmann R, Kuhn A. Macroporous ultramicroelectrodes for improved electroanalytical measurements. Anal. Chem., 2007; 79(2), 533-539.

80. Hermans A, Wightman M. Conical tungsten tips as substrates for the preparation of ultramicroelectrodes. Langmuir, 2006; 22 (25), 10348-10353.4

81. Sachs C, Hildebrand M, Volkening S, Ertl G, Spatiotemporal self-organization in a surface reaction: from the atomic to the mesoscopic scale. Science 2001; 293 (5535) 1635-1638.

82. Zamborini F.P, Crooks, R.M. In-situ electrochemical scanning tunneling microscopy (ECSTM) study of cyanide-induced corrosion of naked and hexadecyl mercaptan-passivated Au(111). Langmuir 1997; 13(2) 122-126.

83. Papadantonakis, K.M., Brunschwig, B.S., Lewis, N.S. Use of Alkane Monolayer Templates To Modify the Structure of Alkyl Ether Monolayers on Highly Ordered Pyrolytic Graphite. Langmuir, 2008; 24, 857–861.

84. Stieg A. Z, Rasool H. I, Gimzewski J.K. A flexible, highly stable electrochemical scanning probe microscope for nanoscale studies at the solid-liquid interface. Review of Scientific Instruments 2008; 79 (10) 103701-7.

85. Stoll E, Marti O. Restoration of scanning-tunneling-microscope data blurred by limited resolution, and hampered by 1/f like noise. Surface science 1987; 181(1-2) 222-229.

86. Eigler D.M, Schweizer E.K. Positioning single atoms with a scanning tunnelling microscope. Nature 1990; 344, 524-526.

87. Reiss G, Bruckl H, Vancea J, Lecheler R, Hastreiter E. Scanning tunneling microscopy on rough surfaces-quantitative image analysis. J. Appl. Phys.1991; 70, 523-525.

88. Endres, F., Borisenko, N., Abedin, S. Z. E., Hayes, R., Atkin, R. The interface ionic liquid(s)/electrode(s): In situ STM and AFM measurements Faraday Discuss., 2012; 154, 221-233.

89. Atkin, R., Abedin, S. Z. E., Hayes, R., Gasparotto, L.H.s., Borisenko, N., Endres, F. AFM and STM Studies on the Surface Interaction of [BMP]TFSA and [EMIm]TFSA Ionic Liquids with Au(111). J. Phys. Chem. C, 2009, 113 (30), 13266–13272.

90. Macpherson, J.V., Unwin, P.R., Combined Scanning Electrochemical–Atomic Force Microscopy. Anal. Chem., 2000, 72 (2), 276–285.

91. Hai, N.T.M., Huynh, T.M.T., Fluegel, A., Mayer, D., Broekmann, P. Adsorption behavior of redox-active suppressor additives: Combined electrochemical and STM studies. Electrochimica Acta. 2011; 56, 7361-7370.

92. Gewirth, A.A., Niece, B.K. Electrochemical Applications of in Situ Scanning Probe Microscopy, Chem. Rev. 1997; 97, 1129–1162.

Reduced Cytotoxicity of Insulin-immobilized CdS Quantum Dots Using PEG as a Spacer

KM Kamruzzaman Selim[1], Zhi-Cai Xing[1], Moon-Jeong Choi[2], Yongmin Chang[3], Haiqing Guo[4], and Inn-Kyu Kang[1]

[1]Department of Polymer Science and Engineering, Kyungpook National University, Daegu 702-701, South Korea

[2]Medical and Biological Engineering, Kyungpook National University, Daegu 702-701, South Korea

[3]Department of Diagnostic Radiology, Kyungpook National University, Dongin-dong, Daegu 700-422, South Korea

[4]College of Chemistry and Molecular Engineering, Peking University, Beijing 100871, China

ABSTRACT

Cytotoxicity is a severe problem for cadmium sulfide nanoparticles (CSNPs) in biological systems. In this study, mercaptoacetic acid-coated CSNPs, typical semiconductor Q-dots, were synthesized in aqueous medium by the arrested precipitation method. Then, amino-terminated polyethylene glycol (PEG) was conjugated to the surface of CSNPs (PCSNPs) in order to introduce amino groups to the surface Finally, insulin was immobilized on the surface of PCSNPs (ICSNPs) to reduce cytotoxicity as well as to enhance cell compatibility. The presence of insulin on the surface of ICSNPs was confirmed by observing infrared absorptions of amide I and II. The mean diameter of ICSNPs as determined by dynamic light scattering was about 38 nm. Human fibroblasts were cultured in the absence and presence of cadmium sulfide nanoparticles to evaluate cytotoxicity and cell compatibility. The results showed that the cytotoxicity of insulin-immobilized cadmium sulfide nanoparticles was significantly suppressed by usage of PEG as a spacer. In addition, cell proliferation was highly facilitated by the addition of ICSNPs. The ICSNPs used in this study will be potentials to be used in bio-imaging applications.

INTRODUCTION

Recently, quantum dots [CdS, CdSe, ZnS, CdTe, etc.] (Q-dots) have attracted tremendous interest as luminescent probes in biological and medical researches due to their unique optical and chemical properties [1]. Compared with traditional dyes and fluorescent proteins used as imaging probes, Q-dots have several advantages, such as tunable emission from visible to infrared wavelengths, broader excitation spectra, high quantum yield of fluorescence, strong brightness, photostability, and high resistance to photobleaching [2,3]. However, the potential applications of Q-dots in biology and medicine have been limited due to their cytotoxic effects [4]. Q-dots contain toxic components such as cadmium (from cadmium chalcogenide-based Q-dots) or lead (from lead chalcogenide-based Q-dots). Cd^{2+} and Pb^{2+} can be released from Q-dots, which would kill the cells [5]. Therefore, to enhance stability, the surface modification of Q-dots is required. For example, biomedical applications require high-quality water soluble and non-toxic Q-dots. So far, numerous surface modifications of Q-dots have been explored, including the attachment of mercaptoacetic acid [6], mercaptopropionic acid [7], mercaptobenzoic acid [8], and biocompatible and chemically functionalizable inorganic shells, such as silica or zinc sulfide [9]. All of these coatings can ensure the water solubility of Q-dots, but they are unable to enhance biocompatibility. Therefore, further coating with suitable water-soluble organic ligand/ biomolecules is necessary to enhance the biocompatibility of Q-dots. To this end, Q-dots have been covalently linked with biorecognition molecules such as biotin [10], folic acid[11], peptides [12], bovine serum albumin [13], transferrin [14], antibodies [15], and DNA [16].

Polyethylene glycol (PEG) and its derivatives have been widely used as biomedical materials, such as drug delivery matrices and scaffolds for tissue engineering, due to their hydrophilicity, high solubility in aqueous and organic solvents, excellent biocompatibility, lack of toxicity and immunogenicity, and ease of excretion from living organisms. Among PEG derivatives, the most important one is amino-terminated PEG [17]. On the other hand, insulin, which reduces blood glucose levels, is often used for treating diabetic patients. However, insulin also acts as a growth factor, inducing cell proliferation[18,19]. It has been previously shown by research groups [18-21] that immobilized insulin stimulates cell growth more actively than free insulin. Therefore, introduction of PEG-insulin conjugate onto the surface of Q-dots through chemical bonding may confer the combined advantage of PEG and insulin. Introduction of PEG onto the surface of nanoparticles protects against unwanted agglomeration, makes them more biocompatible, and decreases their nonspecific intracellular uptake. On the contrary, insulin grafted onto the distal end of the PEG chain can enhance cells growth.

In this study, mercaptoacetic acid-coated cadmium sulfide nanoparticles (CSNPs), typical semiconductor Q-dots, were synthesized in aqueous medium by the arrested precipitation method at room temperature. Then, PEG with amino groups at both ends was reacted with carboxyl groups of CSNPs (PCSNPs) in order to introduce amino groups to the surface as well as to enhance biocompatibility. Finally, insulin was immobilized on the surface of PCSNPs (ICSNPs) to promote cell growth and further enhance biocompatibility. The surface properties of CSNPs and ICSNPs were characterized by X-ray diffraction (XRD), Fourier transform infrared (FT-IR) spectroscopy, transmission electron microphotography (TEM), and dynamic light scattering (DLS). Finally, human fibroblasts were cultured in the presence of nanoparticles to evaluate cell proliferation and cytotoxicity.

EXPERIMENTAL

Preparation of Mercaptoacetic Acid-coated CdS Quantum Dots (CSNPs)

Water-soluble CSNPs were synthesized by following a previously published method [6]. Briefly, carboxyl-stabilized CSNPs were synthesized by arrested precipitation at room temperature in aqueous solution using mercaptoacetic acid as the colloidal stabilizer. Nanocrystals were prepared from a stirred solution of 0.0456 g of $CdCl_2$ (5 mM) in 40 ml of pure water. The pH was lowered to 2 with mercaptoacetic acid and then raised to 7 with 1 N NaOH. The mixture was deaerated by N_2 bubbling for about 30 min, after which 40 ml of freshly prepared 5 mM Na_2S (0.0480 g of Na_2S in 40 ml of water) was added to the mixture with rapid stirring. The solution turned yellow shortly after the sulfide addition due to the formation of CSNPs. CSNPs were separated from reaction by-products (sodium salt) via precipitation by the addition of acetone (4 ml of acetone per milliliter of nanocrystal solution). The precipitate was then isolated by centrifugation and dried in a freeze dryer. The prepared powder CSNPs were finally redispersed in water to obtain a clear colloidal solution with excellent stability (zeta potential, -66.65 mV). The free carboxylic acid groups of the prepared CSNPs are suitable for covalent coupling with the primary amino groups of various biomolecules.

Immobilization of Insulin on the Surface of CSNPs

Immobilization of insulin on CSNPs was performed in two steps. First, CSNPs were reacted with amino-terminated polyethylene glycol (PEG) to introduce amine groups on their surface. For this, CSNPs (0.2 g) were dissolved in aqueous solution (20 ml) containing 1-ethyl-3-(3-dimethylaminopropyl)carbodiimide (EDC) and stirred for 4 h to activate the carboxylic acid groups on the surface. Then, an excess amount of amine-terminated PEG was added to the solution, which was stirred for 24 h to obtain PEG-grafted CSNPs (PCSNPs). An excessive amount of PEG was used to suppress the crosslinking reaction on the surface and keep free amine groups at one end of the PEG chain after the reaction [20]. Prepared PCSNPs were isolated via repeated centrifugation and finally dried in a freeze dryer. In the second step, insulin was immobilized on the surface of PCSNPs as follows: insulin was dissolved in phosphate buffer solution (2 mg/ml, pH 7.4) followed by the addition of a small amount of 0.1 N HCl. Then, 2% w/v water-soluble EDC and NHS were added to the solution, which was incubated at 4°C for 5 h to activate the carboxylic acid groups of the chain. Then, PCSNPs (5 mg/ml) were suspended in phosphate buffer solution (pH 7.4) with vortexing. This PCSNP suspension was mixed with the insulin aqueous solution and stirred gently overnight at room temperature to obtain insulin-immobilized PCSNPs (ICSNPs), as shown in. ICSNPs were isolated by repeated centrifugation and stored in phosphate-buffered saline (PBS) at pH 7. All conjugation reactions, unless otherwise noted, were carried out in the dark under a N_2 ambient environment.

Surface Characterization

Fourier transform infrared (FT-IR) spectra were obtained using a JASCO FT-IR 300E spectrometer (JASCO Inc., Easton, MD, USA) at a resolution of 4 cm^{-1}. Dried samples were ground with KBr powder and compressed into pellets for FT-IR examination. The samples were prepared by dropping the diluted nanoparticles on carbon-coated grids, followed by natural drying; then, the samples were observed by a transmission electron microphotograph (Philips CM 200 TEM; applied operation voltage, 120 kV; Philips Inc, Berlin, Germany. The hydrodynamic diameter and size distribution were determined by DLS by means of a standard laboratory-built light scattering spectrometer equipped with a BI 90 particle sizer (Brookhaven Instruments Corp., Holtsville, NY, USA). It had a vertically polarized incident light of 514.5 nm supplied by an argon ion laser (Lexel laser, model 95; Cambridge Lasers Laboratories Inc., Fremont, CA, USA. To investigate the crystal structure of CSNPs and bare CdS, XRD (RA/FR-571, Enraf Nonius, Deift, The Netherlands was used. The result was also compared with Joint Committee on Powder Diffraction Standards (JCPDS) file no. 10-454 to confirm whether or not any impurity phase exists in the CSNPs. The surface chemical composition was analyzed by electron spectroscopy for chemical analysis (ESCA, ESCA LAB VIG microtech, Mt 500/1, etc., East Grinstead, UK) with MgK α at 1, 253.6 eV and 150

W of power at the anode. A survey scan spectrum was taken, and the surface elemental compositions relative to the carbon were calculated from the peak heights, taking into account atomic sensitivity. The zeta potential is a very useful way of evaluating the stability of any colloidal system. In this study, the zeta potential was measured with a NicompTM 380 Zeta Potential (ZLS, Tokyo, Japan) employing the electrophoretic light scattering technique and using double-distilled water as a diluent.

In vitro Cell Behavior

MRC-5 human fibroblast cells (ATCC CCL, 171) were used in this experiment. Cells were routinely cultured at 37°C in a humidified atmosphere of 5% CO_2 (in air) in a 75-cm^2 flask containing 10 ml of Dulbecco's modified eagle medium (DMEM) supplemented with 10% fetal bovine serum (FBS) and 1% penicillin streptomycin G sodium. The medium was changed every 3 days. For subculture, the cells were washed twice with PBS and incubated with trypsin-ethylenediaminetetraacetic acid (EDTA) solution (0.25% trypsin, 1 mM EDTA) for 10 min at 37°C to detach the cells. The cells were washed twice by centrifugation and resuspended in DMEM media containing quantum dot nanoparticles, including CSNPs, PCSNPs, and ICSNPs (particle concentration, 0.2 mg/ml) for reseeding and growing in culture flasks. The cell density was fixed at 1×10^5 cells/ml. Cell morphologies were observed under a phase contrast microscope (Nikon Eclipse TS100, Tokyo, Japan) at predetermined time intervals.

The proliferation of fibroblasts cultured in the absence and presence of CSNPs, PCSNPs, and ICSNPs was determined by colorimetric immunoassay based on the measurement of 5-bromo-2-deoxyuridine (BrdU), which was incorporated during DNA synthesis [22,23]. BrdU enzyme-linked immunosorbent assay (ELISA; Roche Molecular Biochemicals, Mannheim, Germany) was performed according to the manufacturer's instructions. Briefly, after 48 h of cell culture with CSNPs, PCSNPs, and ICSNPs in 24-well plates, the BrdU-labeling solution was added to each well and allowed to incorporate into the cells for an additional 20 h in a CO_2 incubator at 37°C. Subsequently, the supernatant in each well was removed by pipetting. The cells were then washed twice with PBS and treated with 0.25% trypsin-EDTA (Gibco, Invitrogen, Tulsa, OK, USA) and harvested by centrifugation at 1, 000 rpm for 15 min. The harvested cells were mixed with a FixDenat solution to fix the cells and denature the DNA, followed by further incubation for 30 min. Subsequently, diluted anti-BrdU peroxidase (dilution ratio = 1:100) was added, and the cells were kept at 20°C for 120 min. After the removal of unbound antibody conjugates, 100 µl of substrate solution was added. The resulting mixture was allowed to stand for 20 min, and the reaction was completed by adding 1 M H_2SO4 solution. The solution was then transferred to a 96-well plate and measured within 5 min at 450 nm with a reference wavelength of 690 nm using an ELISA plate reader.

Cytotoxicity

To evaluate the cytotoxicity of Q-dots, the cells were separately cultured in a dish containing CSNPs, PCSNPs and ICSNPs and in a polystyrene culture dish alone. For qualitative observation, Live/Dead fluorescent staining with a LIVE/DEAD Cytotoxicity Kit (Biovision research products, Mountain view, CA 94043 USA) was used. Briefly, fibroblasts (3×10^4 cell/well) were seeded in a microplate with 1 ml of media containing CSNPs, PCSNPs, and ICSNPs (particle concentration = 0.1 mg/ml) without nanoparticles. After 2 and 4 days of incubation, the media were removed and the cells washed gently with PBS. Then, 0.3 ml of staining solution (prepared by mixing calcein-AM and propidium iodide with staining buffer at a concentration specified by Molecular Probes) was added to each well, and the plate was kept in an incubator for 15 min. Then, calcein/propidium iodide solution was removed, and the cells were washed once again with PBS. Finally, cells were viewed using a fluorescence microscope (FV-300, Olympus Co., Tokyo, Japan) coupled with a digital camera (FV-300, Olympus Co.). Live cells show green fluorescence images and dead cells show red images.

Statistical Analysis

The cell viability experiment was performed in triplicate, and the results are expressed as mean ± standard deviation. Student›s t test was employed to assess statistical significant difference of the results.

RESULTS AND DISCUSSION

Characterization of Surface-modified CdS Nanoparticles

The X-ray diffraction spectra of CSNPs and bare CdS are shown in Figure 1. It was observed that the number and positions of peaks of CSNPs (Figure 1a) matched well with those of bare CdS (Figure 1b). The spectrum of CSNPs was further compared with the data of JCPDS file no. 10-454 and was in agreement with that of pure cubic-phase CdS, without signals from $CdCl_2$, NaOH, or other precursor compounds. The three peaks observed in Figure 1a at 2θ values of 26.439°, 43.862°, and 51.389° were found to correspond to the three crystal planes of (111), (220), and (311), indicating that the CSNPs were in cubic phase [24]. Again, the diffraction peaks of CSNPs were somewhat broad compared to those of bare CdS. This broadness was due to reduced particle size and surface defects[25]. Small-sized CSNPs possess a higher surface defect density due to a high surface-to-volume ratio [26]. Moreover, CSNPs possess higher negative zeta potential (ξ = -66.65 mV), indicating excellent stability of the colloidal nanocrystal [27]. The surface modification of CSNPs with insulin was confirmed by FT-IR as displayed in Figure 2. For CSNPs (Figure 2a), two distinctive bands were observed at 1, 559 and 1, 375 cm^{-1}, which originated from the asymmetric and symmetric stretching motion of carboxylate ion (-COO$^-$) [28]. These findings clearly indicate the formation of a co-ordinate bond between the oxygen atom of mercaptoacetic acid and Cd^{+2}. No free carboxylic acid band at 1, 730 to 1700 cm^{-1} due to C=O stretching is observed in capped nanoparticles [29]. The introduction of PEG onto the surface of CSNPs was confirmed by the characteristic peak at 1, 575 cm^{-1}, which can be attributed to a -CH$_2$ bending vibration (Figure 2b). Besides, a peak at 1, 106 cm^{-1} indicated an ether bond (-C-O-) of PEG. Some other peaks were observed at positions of 2, 972, 1, 455, and 1, 375 cm^{-1}, which originated from the PEG chain. This implies that the basic structure of PEG did not change, except for the conversion of a terminal group [17]. After reaction of PEG-immobilized CSNPs (PCSNPs) with insulin, two new peaks at positions around 1, 648 and 1, 540 cm^{-1} were observed based on -CO-NH- (amide I) and

-CO-NH- (amide II) bands, respectively [20] (Figure 2c). These results indicate that insulin was successfully immobilized onto the surface of PCSNPs.

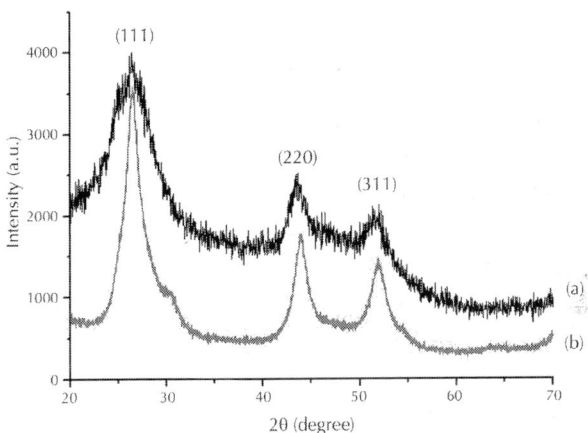

Figure 1: XRD patterns of (a) mercaptoacetic acid-coated CSNPs and (b) bare CdS.

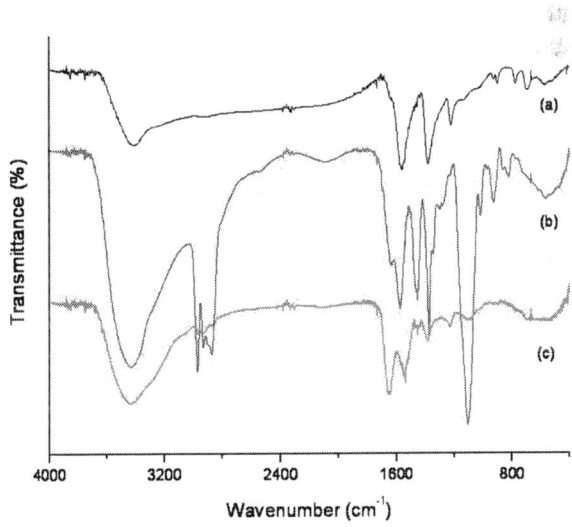

Figure 2: FT-IR spectra of (a) mercaptoacetic acid-coated CSNPs, (b) PCSNPs, and (c) ICSNPs.

Immobilization of insulin onto the surface of CSNPs was further confirmed by ESCA. The chemical compositions of CSNPs, PCSNPs, and ICSNPs, as calculated from the ESCA survey scan spectra, are shown in Table 1. In the case of PCSNPs, the oxygen content (22.09%) and carbon content (62.09%) were higher in comparison to those of CSNPs (oxygen content, 17.35% and carbon content, 30.04%). Furthermore, one new element such as nitrogen (1.59%) was observed on the surface of PCSNPs, indicating the successful immobilization of PEG onto the surface of CSNPs. In the case of ICSNPs, nitrogen content increased from 1.59% to 2.72% and oxygen content increased from 22.09% to 35.81%, indicating the successful immobilization of insulin onto the surface of PCSNPs. TEM images of CSNPs and ICSNPs are shown in Figure 3. It was observed that CSNPs had spherical morphologies with an average diameter of *ca.* 4.5 nm. Due to the small dimensions and high surface energy of the particles, it was easy for them to aggregate as seen in Figure 3a. On the other hand, immobilization of insulin conferred a spherical morphology, thereby reducing the aggregation of particles. The average diameters of the ICSNPs were 13 nm as shown in Figure 3b. Larger diameters and lower aggregation of particles may have resulted from the immobilization of insulin and PEG onto the surface of CSNPs. Figure 4 shows the typical size and size distribution of synthesized CSNPs (Figure4a) and ICSNPs (Figure 4b) as measured by DLS. The average size of CSNPs as determined by DLS was *ca.* 21 nm. On the other hand, the average sizes of ICSNPs were about 38 nm. The size of the particles as determined by DLS was considerably larger than that determined by TEM, most likely because the DLS technique gives the mean hydrodynamic diameter of the core of CSNPs surrounded by organic and solvated layers, which is influenced by the viscosity and concentration of the solution. On the other hand, TEM gives the diameter of the core alone [30].

Table 1: Atomic percent of CSNPs, PCSNPs, and ICSNPs calculated from ESCA survey scan spectra

Sample	Atomic percent (%)				
	C	O	N	S	Cd
CSNP	30.04	17.35	-	16.26	36.35
PCSNP	62.09	22.09	1.59	6.71	7.52

| ICSNP | 56.79 | 35.81 | 2.72 | 1.33 | 3.35 |

Selim *et al. Nanoscale Research Letters* 2011 6:528, doi:10.1186/1556-276X-6-528

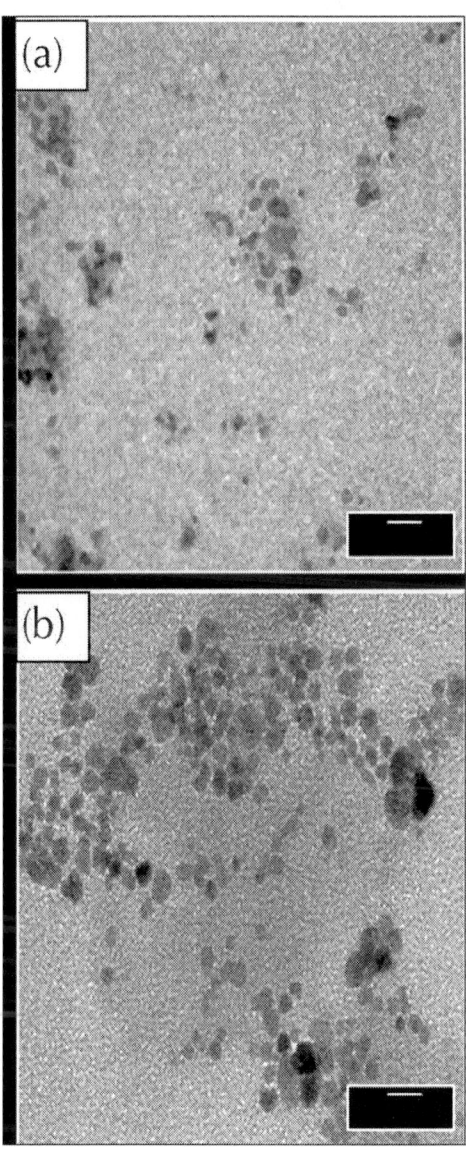

Figure 3: TEM images of (a) mercaptoacetic acid-coated CSNPs and (b) ICS-NPs.

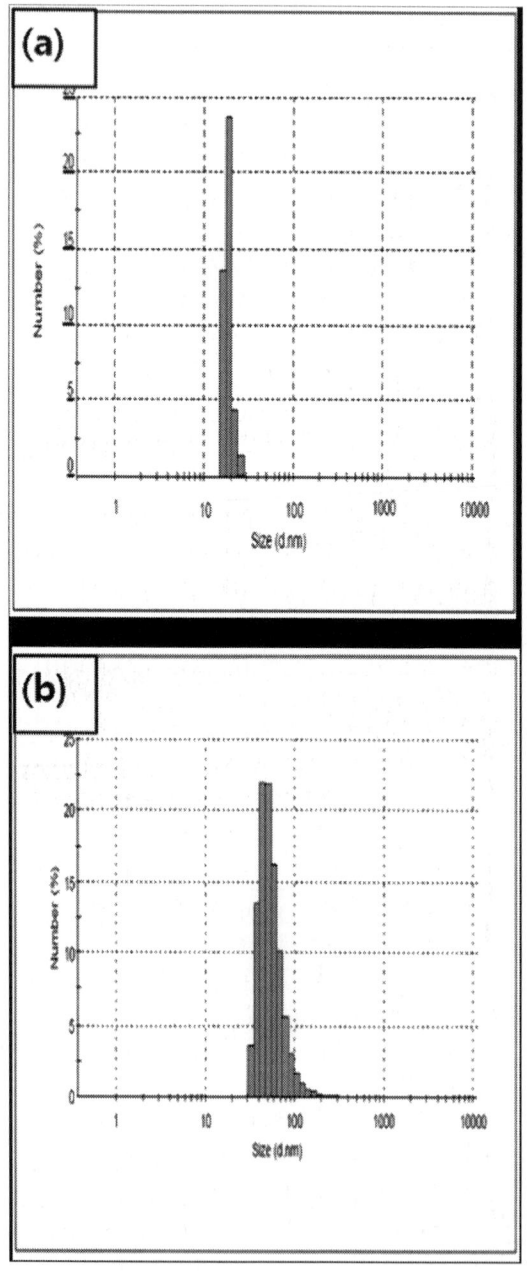

Figure 4: Particle size distributions of (a) mercaptoacetic acid-coated CSNPs and (b) ICSNPs. As measured by dynamic light scattering.

Cell Proliferation

The proliferation of fibroblasts was evaluated by two different methods, BrdU assay and morphological observation. Figure 5 shows the pattern of fibroblast proliferation as measured by BrdU assay after 6 and 24 h of culture in media containing CSNPs, PCSNPs, and ICSNPs. A significant difference in the acceleration of cell growth was not observed after 6 h of culture with CSNPs (Figure6b), PCSNPs (Figure 6c), ICSNPs (Figure 6d), or without nanoparticles (Figure 6a). This could be attributed to the non-interference of particles during the short incubation period. In this case, fibroblasts adhesion occurs due to the influence of FBS-containing media only. However, after 24 h of culture, cell proliferation in media containing ICSNPs was significantly accelerated compared to that in media only However, cell proliferation in media containing PCSNPs or CSNPs was not significantly different from that in media only ($p < 0.004$). Fibroblast proliferation was suppressed in media containing CSNPs. This low cell proliferation was probably due to the negative charge of the surface carboxyl groups on CSNPs [19]. Cell proliferation in media containing PEG-immobilized CSNPs (PCSNPs) was almost the same as that in the control culture dish. This was because the biocompatibility of PEG has already been proven [17]. On the other hand, fibroblast proliferation in media containing ICSNPs was the highest, probably because immobilized insulin molecules sufficiently and continuously stimulate receptors expressed on the plasma membrane surface as well as downstream signal transduction proteins without internalization of ligand-receptor complexes [18]. These results suggest that the binding of immobilized insulin with insulin receptors is essential for the acceleration of the cell proliferation. Similar studies have been reported elsewhere [31,32]. Kim et al.[20] prepared insulin-immobilized polyurethanes and evaluated their interaction with human fibroblasts. As a result, cells were more rapidly proliferated onto insulin-immobilized polyurethanes compared to that on both polyurethane (PU) control and PEO-grafted PU when cultured in the presence of serum. Cell proliferation in the presence of CSNPs, PCSNPs, and ICSNPs and in the absence of nanoparticles was further visualized using an optical microscope. As a result, cell proliferation in the presence of ICSNPs was found to be higher than that in the presence of CSNPs and PCSNPs, as shown in Figure 6.

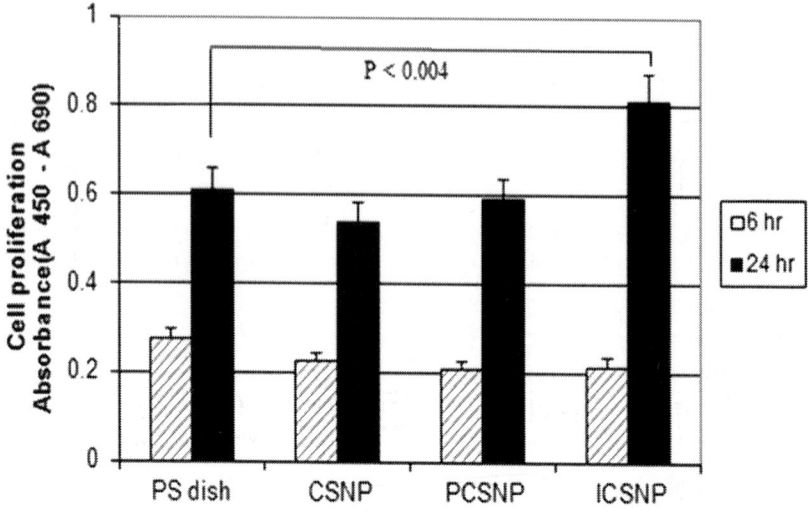

Figure 5: Proliferation of human fibroblasts after 6 and 24 h of incubation. In a dish containing CSNPs, PCSNPs, and ICSNPs and in a polystyrene culture dish, as measured by BrdU assay

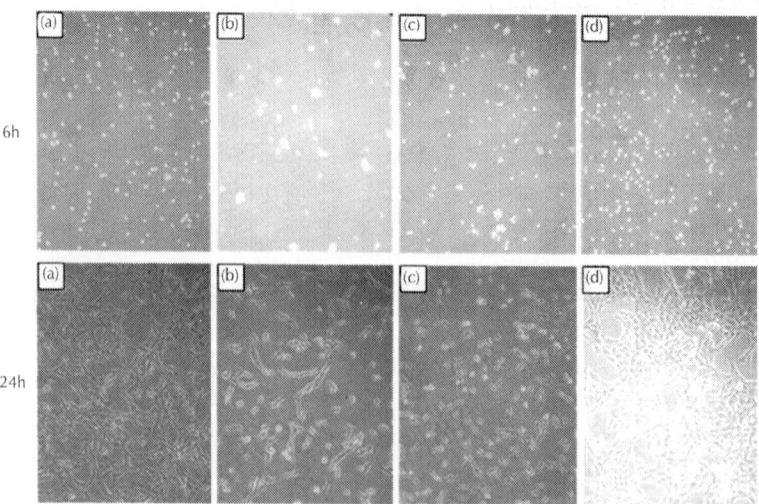

Figure 6: Optical microscopic images of human fibroblasts cultured for 6 and 24 h. In a (a) polystyrene culture dish and in the presence of (b) CSNPs, (c) PCSNPs, and (d) ICSNPs Original magnification is ×200.

Cellular Cytotoxicity

The status of the "Live/Dead" dye-stained fibroblasts cultured in the presence of CSNPs, PCSNPs, and ICSNPs and in the absence of nanoparticles for 2 days is shown in Figure 7. Using this qualitative method, live and dead cells were stained green and red, respectively, under a fluorescence microscope. The color of the cells cultured on a PS dish and with ICSNPs was green, indicating good viability. On the other hand, when cultured with CSNPs and PCSNPs, the green color was partially mixed with red, showing that some parts of the cells were dead. Possible explanations are: (1) toxic cadmium ion (Cd^{+2}) release from CSNPs due to surface oxidation causes cell death [4,33]; (2) reactive oxygen species (ROS) react with cellular biomolecules, resulting in damage, degradation, and finally loss of function [34,35]; (3) nanoparticles are taken up by the cells as a result of endocytosis, which results in disruption of the cell membrane [36]; or (4) weak cell adhesive interactions with CSNPs promote apoptosis (programmed cell death) [36]. Since mercaptoacetic acid is the least solubilizing ligand, and it alone could not protect CSNPs from surface oxidation and diffusion of Cd^{+2} ions from CSNPs over a longer period, it is difficult to make CSNPs biological inert. Therefore, most of the cells died in the presence of CSNPs [4,33,34]. Again, the cell viability of PCSNPs was moderately low due to the surface immobilization conferred by biocompatible PEG, which reduced the release of cadmium ion (Cd^{+2}) and formation of ROS. On the other hand, ICSNPs revealed no cytotoxic effects on cells for up to 2 days. This increased cell viability can be explained by a nutrient effect [36,37]. Besides, the low toxicity of nanoparticles immobilized with insulin may be attributed to the fact that these ligands act as cellular markers that target surface receptors expressed on the cell surface without being internalized. Receptors are highly regulated cell surface proteins that mediate specific interactions between cells and their extracellular milieu, and they are generally localized to the plasma membrane [36]. Based on this explanation, it could be said that the immobilization of biomolecules onto the surface of Q-dots can suppress their toxicity. In this study, PEG and insulin in combination reduced the cytotoxicity of cells.

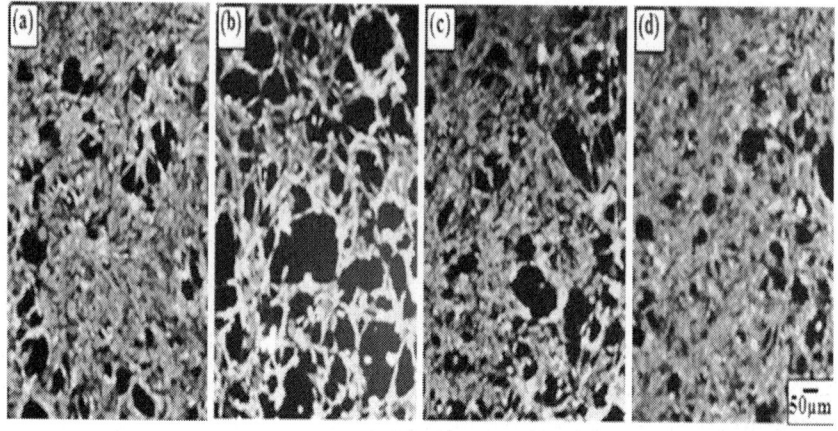

Figure 7: Fluorescence microscopic images of Live/Dead dye-stained fibroblasts cultured for 2 days. In a (a) polystyrene culture dish and in the presence of (b) CSNPs, (c) PCSNPs, and (d) ICSNPs Live and dead cells were stained and visualized in green and red, respectively, under a fluorescence microscope.

CONCLUSIONS

Insulin was immobilized onto the surface of mercaptoacetic acid-coated cadmium sulfide nanoparticles (CSNPs), and confirmation of insulin immobilization was carried out by FT-IR and ESCA. Size distribution of insulin-immobilized CSNPs (ICSNPs) having an average diameter of 13 nm as determined by TEM was narrow. The proliferation of fibroblasts was significantly increased by the presence of ICSNPs. ICSNPs also demonstrated lower cytotoxicity than CSNPs and PCSNPs. The ICSNPs used in this study will have potentials to be used in bio-imaging applications.

AUTHORS' CONTRIBUTIONS

KMK carried out the preparation and immobilization research work. ZCX and MJC participated in the data processing. YC, HG, IKK participated in the design of the study and performed the statistical analysis. All authors read and approved the final manuscript.

ACKNOWLEDGEMENTS

This work was supported by the Korea Ministry of Education, Science and Technology (contract no. 2009-0073282).

REFERENCES

1. Pan J, Feng SS: Targeting and imaging cancer cells by folate-decorated, quantum dots (QDs)-loaded nanoparticles of biodegradable polymers. *Biomaterials* 2009, 30:1176.

2. Alivisatos AP, Gu WW, Larabell C: Quantum dots as cellular probes. *Annual Review of Biomedical Engineering* 2005, 7:55.

3. Michalet X, Pinaud FF, L. Bentolila A, Tsay JM, Doose S, Li JJ: Quantum dots for live cells, *in vivo* imaging and diagnostics. *Science* 2005, 307:538.

4. Derfus AM, Chan WCW, Bhatia SA: Probing the cytotoxicity of semiconductor quantum dots. *Nano Letters* 2004, 4:11.

5. Yu WW, Chang E, Drezek R, Colvin VL: Water-soluble quantum dots for biomedical applications. *Biochemical and Biophysical Research Communications* 2006, 348:781.

6. Chen HM, Huang XF, Xu L, Xu J, Chen KJ, Feng D: Self-assembly and photoluminescence of CdS-mercaptoacetic clusters with internal structures. *Superlattices and Microstructures* 2000, 27:1.

7. Mitchell GP, Mirkin CA, Letsinger RL: Programmed assembly of DNA functionalized quantum dots. *Journal of the American Chemical Society* 1999, 121:8122.

8. Chen CC, Yet CP, Wang HN, Chao CY: Self-assembly of monolayers of cadmium selenide nanocrystals with dual color emission. *Langmuir* 1999, 15:6845.

9. Chen F, Gerion D: Fluorescent CdSe/ZnS nanocrystal-peptide conjugates for long-term, nontoxic imaging and nuclear targeting in living cells. *Nano Letters* 2004, 4:1827.

10. Bruchez M Jr, Moronne M, Gin P, Weiss S, Alivisatos AP: Semiconductor nanocrystals as fluorescent biological labels. *Science* 1998, 28:2013.

11. Chan WCW, Maxwell DJ, Gao X, Bailey RE, Han M, Nie S: Luminescent quantum dots for multiplexed biological detection and imaging. *Current Opinion in Biotechnology* 2002, 13:40.

12. Pinaud F, King D, Moore HP, Weiss S: Bioactivation and cell targeting of semiconductor CdSe/ZnS nanocrystals with phytochelatin-related peptides. *Journal of the American Chemical Society* 2004, 126:6115.

13. Hanaki KI, Momo A, Oku T, Komoto A, Maenosono S, Yamaguchi Y, Yamamoto K:Semiconductor quantum dot/albumin complex is a long-life and highly photostable endosome marker. *Biochemical and Biophysical Research Communications* 2003, 302:496.

14. Chan WCW, Nie S: Quantum dot bioconjugates for ultrasensitive nonisotopic detection. *Science* 1998, 281:2016.

15. Goldman ER, Anderson GP, Tran PT, Mattoussi H, Charles PT, Mauro JM: Conjugation of luminescent quantum dots with antibodies using an engineered adaptor protein to provide new reagents for fluoroimmunoassays. *Analytical Chemistry* 2002, 74:841.

16. Gerion D, Parak WJ, Williams SC, Zanchet D, Micheel CM, Alivisatos AP: Electrophoretic and structural studies of DNA-directed Au nanoparticle groupings. *Journal of the American Chemical Society* 2002, 124:11758.

17. Wang L, Wang S, Bei JZ: Synthesis and characterization of macroinitiator-amino terminated PEG and poly(γ-benzyl-L-glutamate)-PEO-poly(γ-benzyl-L-glutamate) triblock copolymer. *Polymers for Advanced Technologies* 2004, 15:617.

18. Hatakeyama H, Kikuchi A, Yamato M, Okano T: Influence of insulin immobilization to thermoresponsive culture surfaces on cell proliferation and thermally induced cell detachment. *Biomaterials* 2005, 26:5167.

19. Kang IK, Cho SH, Shin DS, Yoon SC: Surface modification of polyhydroxyalkanoate films and their interaction with human fibroblasts. *International Journal of Biological Macromolecules* 2001, 28:205.

20. Kim EJ, Kang IK, Jang MK, Park YB: Preparation of insulin-immobilized polyurethanes and their interaction with human fibroblasts. *Biomaterials* 1998, 19:239.

21. Sasmazel HT, Aday S, Gumusderelioglu M: Insulin and heparin co-immobilized 3D polyester fabrics for the cultivation of fibroblasts in low-serum media. *International Journal of Biological Macromolecules* 2007, 41:338.

22. Maghni K, Nicolescu OM, Martin JG: Suitability of cell metabolic colorimetric assays for assessment of CD4+ T cell proliferation: comparison to 5-bromo-2-deoxyuridine (BrdU) ELISA. *Journal of Immunological Methods* 1999, 223:185.

23. Cui YL, Qi AD, Liu WG, Wang XH, Wang H, Ma DM, Yao KD: Biomimetic surface modification of poly(L-lactic acid) with chitosan and its effects on articular chondrocytes *in vitro*. *Biomaterials* 2003, 24:3859.

24. Smith NV: *X-Ray Powder Data Files*. Philadelphia: American Society for Testing and Materials; 1967.

25. Pucci A, Boccia M, Galembeck F, Leite CAP, Tirelli N, Ruggeri G: Luminescent nanocomposites containing CdS nanoparticles dispersed into vinyl alcohol based polymers. *Reactive and Functional Polymers* 2008, 68:1144.

26. Wang Q, Kuo YC, Wang Y, Shin G, Ruengruglikit C, Huang Q: Luminescent properties of water-soluble denatured bovine serum albumin-coated CdTe quantum dots. *The Journal of Physical Chemistry B* 2006, 110:16860.

27. ASTM Designation F316-86: Standard test methods for pore size characteristics of membrane filters by bubble point and mean flow pore test. 752-757.

28. Kuo YC, Wang Q, Ruengruglikit C, Yu H, Huang Q: Antibody-conjugated CdTe quantum dots for *Escherichia coli* detection. *The Journal of Physical Chemistry C* 2008, 112:4818.

29. Chowdhury PS, Ghosha P, Patra A: Study of photophysical properties of capped CdS nanocrystals. *Journal of Luminescence* 2007, 124:327.

30. Selim KMK, Lee JH, Kim SJ, Xing Z, Chang Y, Guo H, Kang IK: Surface modification of magnetites using maltotrionic acid and folic acid for molecular imaging. *Macromolecular Research* 2006, 14:646.

31. Chen G, Ito Y, Imanishi Y: Mitogenic activities of water-soluble and -insoluble insulin conjugates. *Bioconjugate Chemistry* 1997, 8:106.

32. Li JS, Ito Y, Zheng J, Takahashi T, Imanishi Y: Enhancement of artificial juxtacrine stimulation of insulin by co-immobilization with adhesion factors. *Journal of Biomedical Materials Research Part A* 1997, 37:190.

33. Selim KMK, Xing ZC, Guo H, Kang IK: Immobilization of lactobionic acid on the surface of cadmium sulfide nanoparticles and their interaction with hepatocytes. *Journal of Materials Science: Materials in Medicine* 2009, 20:1945.

34. Selim KMK, Guo H, Kang IK: Albumin-conjugated cadmium sulfide nanoparticles and their interaction with KB cells. *Macromolecular Research* 2009, 17:403.

35. Maysinger D, Lovric J, Eisenberg A, Savic R: Fate of micelles and quantum dots in cells. *European Journal of Pharmaceutics and Biopharmaceutics* 2007, 65:270.

36. Gupta AK, Berry C, Gupta M, Curtis A: Receptor-mediated targeting of magnetic nanoparticles using insulin as a surface ligand to prevent endocytosis. *IEEE Transactions on Nanobioscience* 2003, 2:255.

37. Fischer D, Li YX, Ahlemeyer B, Krieglstein J, Kissel T: *In vitro* cytotoxicity testing of polycations: influence of polymer structure on cell viability and hemolysis. *Biomaterials* 2003, 24:1121.

Chapter 9

Influence of Helium-Ion Bombardment on the Optical Properties of Zno Nanorods/P-Gan Light-Emitting Diodes

Naveed ul Hassan Alvi[1], Sajjad Hussain[1], Jen Jensen[2],
Omer Nur[1], and Magnus Willander[1]

[1]Department of Science and Technology (ITN), Campus Norrköping, Linköping University, 60174 Norrköping, Sweden

[2]Department of Physics, Chemistry and Biology, Linköping University, 58183, Linköping, Sweden

ABSTRACT

Light-emitting diodes (LEDs) based on zinc oxide (ZnO) nanorods grown by vapor-liquid-solid catalytic growth method were irradiated with 2-MeV helium (He^+) ions. The fabricated LEDs were irradiated with fluencies of approximately 2×10^{13} ions/cm^2 and approximately 4×10^{13} ions/cm^2. Scanning electron microscopy images showed that the morphology of the irradiated samples is not changed. The as-grown and He$^+$-irradiated LEDs showed rectifying behavior with the same I-V characteristics. Photoluminescence (PL) measurements showed that there is a blue shift of approximately 0.0347 and 0.082 eV in the near-band emission (free exciton) and green emission of the irradiated ZnO nanorods, respectively. It was also observed that the PL intensity of the near-band emission was decreased after irradiation of the samples. The electroluminescence (EL) measurements of the fabricated LEDs showed that there is a blue shift of 0.125 eV in the broad green emission after irradiation and the EL intensity of violet emission approximately centered at 398 nm nearly disappeared after irradiations. The color-rendering properties show a small decrease in the color-rendering indices of 3% after 2 MeV He$^+$ ions irradiation.

INTRODUCTION

Zinc oxide (ZnO) with bad gap of 3.37 eV has very attractive properties to play a major role in nanoscale electronic and optoelectronic devices. Its excellent properties combined with the easiness of growing it in the nanostructure form have made it one of the most attractive and versatile semiconductor [1,2]. It has both semiconducting and piezoelectric properties and in addition it is biocompatible and bio-safe. ZnO possesses deep level emission (DLE) bands emitting all the colors in the visible region and has good color-rendering properties [3-5].

Among all of the known oxide semiconductors, ZnO nanorods (NRs) are the best choice for intrinsic white light emission due to their easy growth via chemical as well as other physical vapor-phase approaches [6]. ZnO NRs with small footprint and large surface area to volume ratio are good candidates for heterojunction light-emitting diodes (LEDs) as compared to thin films. This is due to the fact that the stress/strain due to lattice mismatch can easily be released for nanorods when compared to thin films. In addition, a general property of NRs-based LEDs is that each nanorod can act as a wave guide, minimizing side scattering of light, thus enhancing light emission and extraction efficiency [7]. It is still a challenge to achieve a reproducible, high quality p-type epitaxial technology for ZnO. This hinders the progress of ZnO homojunction LEDs. The alternative way is to grow n-type ZnO nanostructures on top of other p-type substrates to make heterojunctions [8-10]. The close lattice match is the main factor that can influence the optical and electrical properties of heterojunctions. The p-GaN as a substrate is a good candidate that has small lattice mismatch with ZnO. There is only few growth reports of n-ZnO nanorods on p-GaN, and on white LEDs based on them available in the literature, e.g., [11-13].

The properties of a material can be changes by irradiation of that material with energetic particles such as electrons or ions that normally gives rise to formation of defects in the target material [14]. An important consideration for space and nuclear applications of ZnO nanostructures based LEDs is that these LEDs should be reliable to withstand and to operate in radiation hard conditions. In this paper, we have investigated the effect of 2 MeV He+ ion irradiation on the optical properties of ZnO nanorod-based LEDs. There are only few reports about the effect of high-energy electrons irradiation that has been reported for ZnO [15], GaN [16,17], and SiC [18]. Effect of ion and electron irradiation on the properties of nanostructured materials has been studied [14]. Much less data are available regarding the effect of heavier particles (such as He+ ions) on the physical properties of ZnO nanostructures [19]. The irradiation changes the amount of defects in ZnO nanorods resulting in changes in the optical properties of ZnO. The effect of the ion irradiation of complete device, like a LED, has not been studied so much.

EXPERIMENTAL DETAILS

ZnO NRs were grown on p-GaN substrates by the vapor-liquid-solid (VLS) mechanism. Gold nano particles are used as catalyst for the growth. A thin film of gold with thickness of approximately 4 nm was deposited on the substrates in a low-vacuum metallization chamber. In this method, pure zinc powder (99.9%) is used as the source material. The pure zinc powder was placed in a quartz tube and the substrates are placed on the boat at the downstream side of the gas flow. The substrates are placed at a distance of 1 to 2 cm away from the zinc powder. The mixture of the argon and oxygen gases with a ratio of 8:1 is introduced in the quartz tube. Argon was applied as a carrying gas and oxygen was as reactant gas [20]. The growth temperature was approximately 680°C.

After the growth of the ZnO NRs on p-GaN substrates, three samples were used to process the LEDs. Pt/Ni/Au alloy was used to form ohmic contact to the p-GaN substrate. The thickness of the Pt, the Ni, and the Au layers were 20, 30, and 80 nm, respectively. The sample was annealed at 350°C for 1 min in flowing argon gas atmosphere. This alloy gives a minimum specific contact resistance of $5.1 \times 10^{-4}\,\Omega$ cm^{-2} [21]. After that an insulating photo-resist layer was spun coated on the ZnO NRs to fill the gaps between the NRs to isolate electrical contacts on the ZnO NRs from reaching the p-type substrate and helps to prevent the carrier cross talk among the nanorods. To form the top contacts, the tips of the ZnO NRs were exposed by using plasma ion etching after the deposition of the insulating photo-resist. Non-alloyed Pt/Al metal system was used to form the ohmic contacts to the ZnO NRs. The thickness of Pt and Al layers were 50 and 60 nm, respectively. This contact gives a minimum specific contact resistance of $1.2 \times 10^{-5}\,\Omega\,cm^{-2}$ [22]. The diameter of the top contact was about 0.58 mm. Two of these fabricated LEDs and two GaN substrates with the as-grown ZnO NRs were are irradiated by using 2.0 MeV He^{+} ions with fluences of approximately 2×10^{13} ions cm^{-2} and approximately 4×10^{13} ions cm^{-2}. All the samples were irradiated at room temperature under normal incidence at the Tandem Laboratory, Uppsala University Sweden. The ion beam flux was at about 6×10^{10} ions $s^{-1}\,cm^{-2}$ and the beam spot was roughly 1 cm^{2}. The projected ion range is calculated by the SRIM2008 code [23] to be 4.9 μm assuming a density for ZnO of 5.6 g/cm^{3}, so the major part of the whole ZnO nanorods were influence by the beam. The irradiated ZnO nanorods are used for PL measurements and the devices were used to measure the IV and the electroluminescence (EL) measurements. The PL measurements were performed at room temperature. Laser lines with a wavelength of 266 nm from a diode laser pumped resonant frequency doubling unit (MBD266) were used as an excitation source. The EL measurements of the fabricated LEDs were performed by using a photomultiplier detector at room temperature. The spectra were measured from the top contacts of the LEDs by detecting the light escaping from the edge of the top contact electrode.

RESULTS AND DISCUSSIONS

The morphology and size distribution of the as-grown ZnO nanorods were investigated by using JEOLJSM-6301F SEM. The top SEM view of as-grown and irradiated ZnO nanorods are shown in Figure 1a, b, c, respectively. The ZnO structures had a uniaxial orientation of <0001> perpendicular to the substrates. The epitaxial growth of the ZnO nanorods with respect to the p-GaN substrates, forming n-ZnO-(nanostructures)/p-GaN p-n heterojunction. From the SEM images, the mean diameters of our as-grown ZnO nanorods was approximately 450 nm. The approximate length of ZnO nanorods was 3 μm.

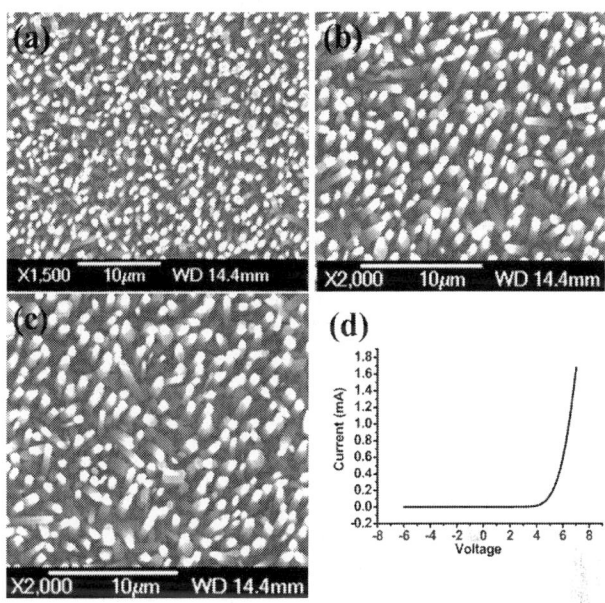

Figure 1: SEM image of ZnO nanorods on p-GaN substrate. (a) As grown, (b) after irradiation with fluency of approximately 2×10^{13}ions/cm^2, (c) after irradiation with fluency of approximately 4×10^{13}ions/cm^2, and (d) typical I-V characteristics for the fabricated LEDs.

The I-V characterization of the fabricated LED is shown in Figure 1d. The irradiated and non irradiated LEDs have same I-V curves. The I-V curve shows rectifying behavior as expected from these LEDs. It indicates clearly that reasonable p-n heterojunctions characteristics were achieved. The turn-on voltage of these heterojunctions LEDs is around 3 V.

Figure 2a, b, c shows the photoluminescence of the as-grown and irradiated ZnO NRs. Figure 3ashows the PL spectrum for as-grown ZnO NRs. The band-edge emission and the DLE peaks are observed at approximately 380 nm (3.26 eV) and 530 nm (2.33 eV). The observation of the band-edge emissions at 380 nm (3.26 eV) are attributed to the first and second longitudinal optical (LO) phonon replica. This is consistent with the LO-phonon energy of 72 meV for ZnO. Their accurance shows that our as-grown ZnO structures are of good crystalline quality [24]. The green emission centered at 2.33 eV (530 nm) is attributed to recombination between the bottom of the conduction band to the O_i energy level and it is approximately agreed with reported data for the transition energy from bottom of the conduction band to O_i energy level (approximately 2.28 eV) [25,26].

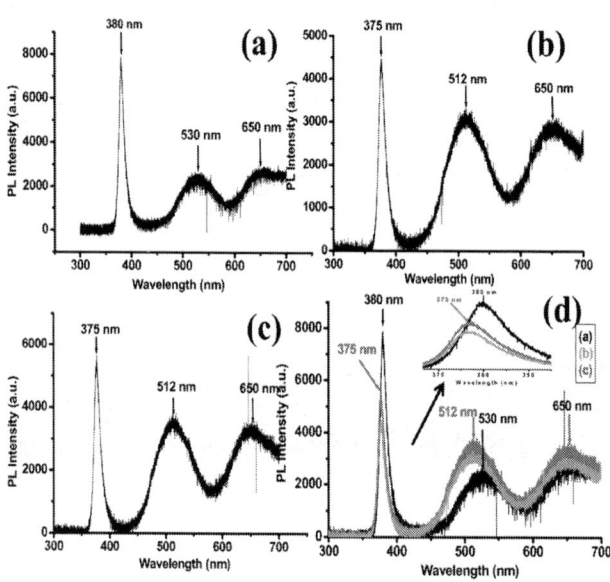

Figure 2: Room-temperature photoluminescence spectrum for ZnO nanorods. (a) As grown, (b) after irradiation with fluency of approximately 2×10^{13} ions/cm^2, (c) after irradiation with fluency of approximately 4×10^{13} ions/cm^2, and (d) the PL spectra of all the samples together for comparison.

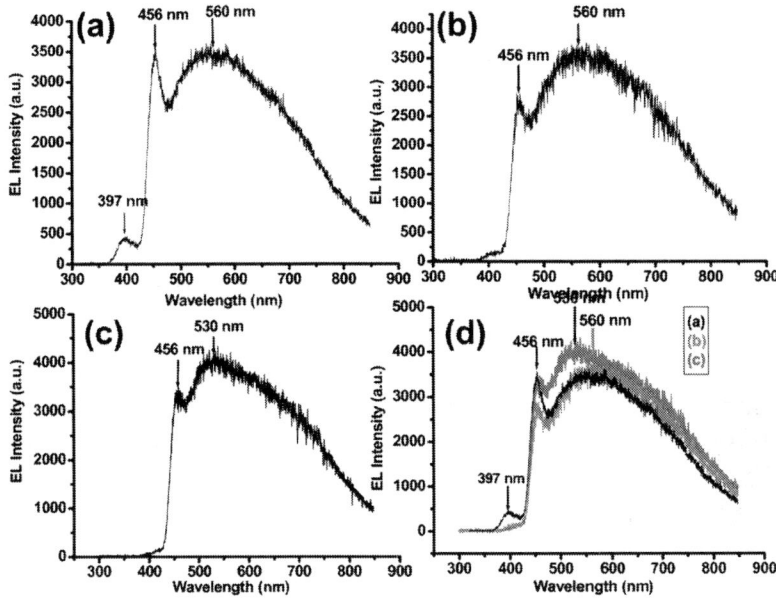

Figure 3: Displays the electroluminescence spectra for n-ZnO nanorods/p-GaN LEDs, (a) As grown, (b) after irradiation with fluency of approximately 2 × 10^{13} ions/cm^2, (c) after irradiation with fluency of approximately 4 × 10^{13} ions/cm^2, and (d) the EL spectra of all the LEDs together for comparison.

Figure 2b, c shows the PL spectrum for irradiated ZnO nanorods with fluencies of approximately 2 × 10^{13} ions/cm^2 and approximately 4 × 10^{13} ions/cm^2, respectively. The band-edge emission and the DLE peaks are observed at approximately 375 nm (3.30 eV), and 512 nm (2.42 eV). The green peak centered at 2.42 eV (512 nm) is attributed to recombination between the conduction band energy level to the V$_O$ energy level and it is approximately consistent with the transition energy from Zn$_i$energy level to V$_O$ energy level (approximately at 2.47 eV). A comparison of the PL spectra of the ZnO NRs before and after the He$^+$ ion irradiation is shown in Figure 2d. It shows that the PL intensity of the near-band-edge decrease and the PL intensity of the deep level emission increase after the He$^+$ion irradiation. The decrease in the UV emission is due to degradation in the crystalline quality after the He$^+$ ion irradiation. The enhancement in the green emission is due to the increase of radiative defects such as oxygen vacancies [27].

It is also observed that there is a blue shift of about 0.0347 eV in the near-band-edge emission after the He$^+$ ion irradiation. The blue shift in the excitonic emission peak can be attributed to the presence of homogeneous compressive strain produced by the He$^+$ ion irradiation. Compressive strain increases the band gap and affects the optical properties of the materials [28]. There is also a blue shift of 0.082 eV in the green emission after He$^+$ irradiation. This blue shift is due to the fact that the He$^+$ ion irradiation increases the oxygen vacancy defects as compared to oxygen interstitials defects in the ZnO. The green emission is the superposition of the emissions due to oxygen vacancies and oxygen interstitials. The oxygen interstitials have lower energy (2.28 eV) as compared to oxygen vacancies (2.47 eV). The as-grown ZnO nanorods have green emission approximately centered at 2.33 eV and it is very close to energy of oxygen vacancies but after He$^+$ irradiation oxygen vacancies defects increase and irradiated ZnO nanorods have green emission is centered at approximately 2.42 eV and it is very close to energy of oxygen vacancies.

The orange-red emission peak is centered at 650 nm (1.90 eV). This orange-red emission can be attributed to the transition from zinc interstitial (Zn_i) to oxygen interstitial (O_i) defect levels in ZnO[26]. By using the full potential linear muffin-tin orbital method, which explains that the position of the O_i level is located approximately at 2.28 eV below the conduction band and the Zn_i level is theoretically located at 0.22 eV below the conduction band. Therefore the transition energy from Zn_i to O_i levels is approximately 2.06 eV [26]. This value agrees approximately with the experimental (EL) peak centered at 1.90 eV.

Figure 3a shows the EL spectra of as-grown ZnO nanorods. The violet, violet-blue, and green emissions are observed and are centered approximately at 397 nm (3.12 eV), 456 nm (2.71 eV) and 560 nm (2.21 eV), respectively. It was reported that the violet emission from undoped ZnO nanorods corresponds to zinc interstitials (Zn_i) [26]. The violet peak is centered at 3.12 eV (397 nm) and it agrees well with the transition energy from the valence band to the Zn_i level in ZnO (approximately 3.1 eV). The violet-blue peak centered at 2.71 eV (456 nm) is attributed to recombination between the Zn_i energy level to the V_{Zn} energy level and it is approximately agreed with the transition energy from Zn_i energy level to V_{Zn} energy level (approximately 2.84 eV). There is a difference of 0.13 eV. Maybe this difference is due to the effect of GaN substrate as GaN also emits blue light. As the violet-blue peak is not observed in the PL spectra this may support the argument that the blue emission is from GaN substrate. The green emission centered at 560 nm (2.21 eV) is attributed to oxygen interstitials and is discussed above. The green emission in the EL spectra is red shifted as compared to the PL spectra. It may be due to the heating effect of the device.

Figure 3b, c shows the EL spectra of He$^+$ ion irradiated LEDs. The EL spectra of irradiated LEDs shows that the violet peak centered at 397 nm (3.12 eV) almost disappeared after He$^+$ ion irradiation. It disappears due to the poor crystalline quality after irradiation. The green emission EL peak is blue shifted about 0.125 eV. The reason of this blue shift in the green peak is discussed above. The blue emission looks stable as it is from the substrate.

Figure 4a, b, c shows the CIE 1931 color space chromaticity diagram in the (x, y) coordinates system. The chromaticity coordinates of the non radiated and He$^+$ ion radiated with fluencies of approximately 2 × 10^{13} ions/cm^2 and approximately 4 × 10^{13} ions/cm^2 LEDs are (0.3557, 0.3805), (0.3735, 4020), and (03594, 0.3983) with correlated color temperatures (CCTs) of 4, 745, 4, 345, and 4, 708 K, respectively. It seems that the chromaticity coordinates for non radiated LED are close to Planckian locus (about 2 Mac-Adam ellipse away) and can be called white light according to the US standard ANSI_ANSLG C78, 377-2008 for the solid-state light sources which determines that the distance of 3 Mac-Adam ellipses form Planckian locus can be allowed for white light. The He+ ion-radiated LEDs are slightly decreased from the Planckian Locus and are very close to white light. They are about 4 Mac-Adam ellipses away from the Planckian Locus. The color-rendering indices are 92, 90, and 89 for the non-radiated and He$^+$ ion radiated with fluencies of approximately 2 × 10^{13} ions/cm^2 and approximately 4 × 10^{13} ions/cm^2 LEDs, respectively. It shows that there is only a small effect (3%) on the optical properties. So this might be a good observation for space and nuclear applications. The 2-MeV ions only have a minor effect on optical properties of the LED devise after irradiation with moderate fluences.

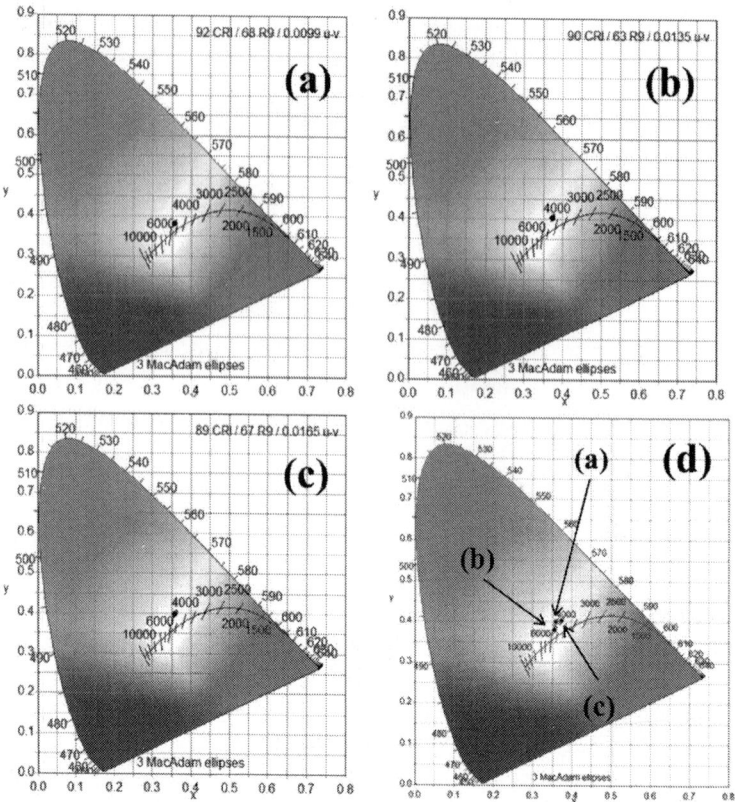

Figure 4: Shows the CIE 1931 x, y chromaticity space. Showing the chromaticity coordinates of LEDs under forward bias for ZnO NRs/p-GaN LEDs, (a) as-grown ZnO NRs, (b) after irradiation with fluency of approximately 2×10^{13} ions/cm², (c) after irradiation with fluency of approximately 4×10^{13} ions/cm², and (d) all together for comparison.

CONCLUSIONS

In summary, the influence of He⁺ ion irradiation on the optical properties of ZnO nanorods has been investigated as a possible candidate for space applications of ZnO nanorods based LEDs. The PL investigations show that the irradiation has influence the high-energy defects in the ZnO, especially the defects responsible for UV, violet and green emission in ZnO. Due to this influence crystallinity in ZnO decreases and as a result PL intensity of ultraviolet (UV) emission decreases. A blue shift in UV and green emission was found. The EL spectra reveal the same blue shift in the green emission as observed from the PL spectra. The irradiation has a minor effect on the color-rendering properties of the LEDs. It decreases the color-rendering indices from 92 to 89.

AUTHORS' CONTRIBUTIONS

All authors contributed equally. All authors read and approved the final manuscript.

REFERENCES

1.   Duan X, Huang Y, Cui Y, Wang J, Lieber CM: Indium phosphide nanowires as building blocks for nanoscale electronic and optoelectronic devices. *Nature (London)* 2001, 409:66-69.

2.   Huang MH, Mao S, Feick H, Yan H, Wu Y, Kind H, Weber E, Russo R, Yang P: Room-temperature ultraviolet nanowire nanolasers. *Science* 2001, 292:1897-1899.

3.   Willander M, Nur O, Zhao QX, Yang LL, Lorenz M, Cao BQ, Zúñiga Pérez J, Czekalla C, Zimmermann G, Grundmann M, Bakin A, Behrends A, Al-Suleiman M, El-Shaer A, Che Mofor A, Postels B, Waag A, Boukos N, Travlos A, Kwack HS, Guinard J, Le Si Dang D: Zinc oxide nanorod based photonic devices: recent progress in growth, light emitting diodes and lasers. *Nanotechnology* 2009, 20:332001.

4.   Wang ZL: Nanostructures of zinc oxide. *Materials Today* 2004, 7:26-33.

5.   Klingshirn C: ZnO: from basics towards applications. *Physica Status Solidi (b)* 2007, 244:3027-3073.

6.   Djurisic AB, Leung YH, Tam KH, Hsu YF, Ding L, Ge WK, Zhong YC, Wong KS, Chan WK, Tam HL, Cheah KW, Kwok WM, Phillips DL: Defect emissions in ZnO nanostructures. *Nanotechnology* 2007, 18:095702.

7.   Lai E, Kim W, Yang P: Vertical nanowire array-based light emitting diodes. *Nano Research* 2008, 1:123-128.

8. Alivov Ya I, Van Nostrand JE, Look DC, Chukichev MV, Ataev BM: Observation of 430 nm electroluminescence from ZnO/GaN heterojunction light-emitting diodes. *Applied Physics Letters* 2003, 83:2943-2945.

9. Chen PL, Ma XY, Yang DR: Ultraviolet electroluminescence from ZnO/p-Si heterojunctions. *Journal of Applied Physics* 2007, 101:053103-053106.

10. Ohta H, Kawamura K, Orita M, Hirano M, Sarukura N, Hosono H: Current injection emission from a transparent p-n junction composed of p-SrCu2O2/n-ZnO. *Applied Physics Letters* 2000, 77:475-477.

11. Yuen C, Yu SF, Lau SP, Rusli , Chen TP: Fabrication of n-ZnO:Al/p-SiC(4H) heterojunction light-emitting diodes by filtered cathodic vacuum arc technique. *Applied Physics Letters* 2005, 86:241111-241113.

12. Mofor AC, Bakin A, Chejarla U, Schlenker E, El-Shaer A, Wagner G, Boukos N, Travlos A, Waag A: Fabrication of ZnO nanorod-based p-n heterojunction on SiC substrate. *Superlattices and Microstructures* 2007, 42:415-420.

13. Lee SD, Kim Y-S, Yi M-Su, Choi J-Y, Kim S-W: Morphology control and electroluminescence of ZnO nanorod/GaN heterojunctions prepared using aqueous solution. *Journal of Physical Chemistry C* 2009, 113:8954-8958.

14. Krasheninnikov AV, Nordlund K: Ion and electron irradiation-induced effects in nanostructured materials. *Journal of Applied Physics* 2010, 107:071301-071400.

15. Look DC, Reynolds DC, Hemsky JW, Jones RL, Sizelove JR: Production and annealing of electron irradiation damage in ZnO. *Applied Physics Letters* 1999, 75:811-813.

16. Fang ZQ, Look DC, Kim W, Fan Z, Botchkarev A, Morkoc H: Deep centers in n-GaN grown by reactive molecular beam epitaxy. *Applied Physics Letters* 1998, 72:2277-2229.

17. Goodman SA, Auret FD, Legodi MJ, Gibart P, Beaumont B: Characterization of electron-irradiated n-GaN. *Applied Physics Letters* 2000, 78:3815-3817.

18. Hallen A, Henry A, Pellegrino P, Svensson BG, Åberg D: Ion implantation induced defects in epitaxial 4H-SiC. *Materials Science and Engineering B* 1999, 378:61-62.

19. Hayes M, Auret FD, Janse van Rensburg PJ, Nel JM, Wesch W, Wendler E: Electrical characterization of He+ irradiated n-ZnO. *Physica Status Solidi (b)* 2007, 244:1544-1548.

20. Fang T-H, Chang W-J, Water W, Lee C-C: Effect of gas concentration on structural and optical characteristics of ZnO nanorods. *Physica E* 2010, 42:2139-2142.

21. Jang J-S, Chang I-S, Kim H-K, Seong T-Y, Lee S, Park S-J: Low-resistance Pt/Ni/Au ohmic contacts to p-type GaN. *Applied Physics Letter* 1999, 74:70-72.

22. Kim H-K, Kim K-K, Park S-J, Seong T-Y, Adesida I: Formation of low resistance nonalloyed Al/Pt ohmic contacts on n-type ZnO epitaxial layer. *Journal of Applied Physics* 2003, 94:4225-4227.

23. SRIM (Stopping and Range of Ion in Matter) Software [http://www.srim.org] *webcite* Version 2008

24. Bekeny C, Kreye M, Waag A: Origin of the near-band-edge photoluminescence emission in aqueous chemically grown ZnO nanorods. *Journal of Applied Physics* 2006, 100:104317-104320.

25. Klason P, Børseth TM, Zhao QX, Svensson BG, Kuznetsov AY, Bergman PJ, Willander M:Temperature dependence and decay times of zinc and oxygen vacancy related photoluminescence bands in zinc oxide. *Solid State Communications* 2008, 145:321-326.

26. Ahn CH, Kim YY, Kim DC, Mohanta SK, Cho HK: A comparative analysis of deep level emission in ZnO layers deposited by various methods. *Journal of Applied Physics* 2009, 105:089902-089901.

27. Liao L, Lu HB, Li JC, Liu C, Fu DJ, Liu YL: The sensitivity of gas sensor based on single ZnO nanowire modulated by helium ion radiation. *Applied Physics Letters* 2007, 91:173110-173112.

28. Jain S, Willander M, Overstraeten RV: *Compound semiconductors strained layers and devices.*Norwell: Kluwer Academic; 2000.

Chapter 10

Theoretical Investigation of Electronic and Optical Properties of Si/SiGe Quantum Cascade Structures

Khadidja Zellat[1], Belabbes Soudini[1], and Salah Mohamed Ait Cheikh[2]

[1]Applied Materials Laboratory (AML), University of Sidi Bel Abbès, Sidi Bel Abbès, Algeria

[2]Laboratoire de Dispositifs de Communication Conversion Photovoltaique, Ecole Nationale Polytechnique, Algiers, Algeria

ABSTRACT

This paper reviews the basic properties of the SiGe alloy, presents some new results on its electronic and optical properties, and discusses the approach that has been followed to model quantum wells containing SiGe layers for applications in quantum cascade lasers. The shape of the confining potential, the subband energies and their eigen envelope wave functions are calculated by solving a one-dimensional

Schrödinger equation. The calculations of optical parameters are used to optimize the Si/SiGe quantum cascade structures. Our results are found to be in good agreement with other calculations.

INTRODUCTION

A quantum cascade laser (QCLs) is a specific type of semiconductor laser that operates through principles of quantum mechanics. Already theoretically predicted in 1971 [1], QCLs had not been realized until 1994 at Bell Laboratories [2]. They have many advantages over other types of semiconductors' lasers. Some of these advantages include precise tuning from one wavelength to another, higher optical power, continuous wave operation and the ability to produce light in the terahertz range of the spectrum [3-6].

From the physical point of view, the unipolarity of a QC laser indicates that electrons are solely responsible for releasing energy in the form of photons. These electrons transition from one quantum energy state to another within a layer, or group of layers, of semiconductor material releasing energy in the form of photons during their descent. The binding energy necessary to pull these electrons away from the Coulombic force of the nucleus in the atom is related to the extremely thin semiconductor layers. A property of quantum mechanics known as quantum confinement occurs when the electrons are trapped within a thin semiconductor quantum well layer. These electrons can freely move in only two directions within the plane of the thin layer. In this case quantum confinement, which leads to discrete energy levels that electrons can occupy in a material smaller than the de Broglie wavelength, occurs in only one dimension due to the quantum well structure [7]. Unlike the earliest form of semiconductor lasers where the energy bandgap determines the wavelength of the light emitted, with QC lasers the thickness of the layers determines the wavelength. This is a critically important property of QC lasers because it allows them to be tuned to a desired frequency through bandgap engineering [8]. Technologically speaking, this laser type is grown by epitaxial method such as molecular beam epitaxy (MBE) [9]. Layers of different semiconductor materials each only a few atomic layers thin are deposited onto a thin slice of a semiconductor crystal. In order to optimize the electronic wave functions with respect to energy and probability distribution, we have to choose the sequence of the layers, their width and materials.

It was established that these unipolar intersubband lasers might be realized not only in III-V semiconductors [10-13] but also in IV-IV structures [14-16]. Among semiconductors, the covalent semiconductors Si and Ge have been studied extensively both theoretically and experimentally [17,18]. Group IV semiconductors alloys like Si-Ge, have the immense potential for technological applications whose include the optoelectronic devices [19-21]. By using intersubband transitions within the same band, one can circumvent the main obstacle to silicon-based lasers, the indirect band gap. The large band offsets in the valence band of pseudomorphic SiGe layers on Si substrates imply a quantum cascade scheme with hole subbands.

In our previous publication [22], we were interested to the investigation of the structural, electronic and optical properties of Si, Ge, and $Si_{1-x}Ge_x$ for different compositions using the full-potential linear muffin-tin orbital (FP-LMTO) method augmented by a plane-wave basis (PLW), implemented in Lmtar code [23-25]. All the obtained results showed that the weakly strained G-rich SiGe layers possess very promising properties for both electronic and optical applications.

The aim of the present work is to provide a consistent and complete set of electronic and optical parameters of the Si/SiGe quantum well. The obtained results are going to be of use to a good understanding of the quantum phenomena of these devices. The second objective concerns the way which allows us to optimize the intrinsic parameters of the Si/SiGe quantum cascade structure.

METHOD OF CALCULATIONS

Much theoretical work has been done to accompany the rapid experimental developments of QCLs as well as to better explain the design considerations of intersubband lasers. These include Monte Carlo simulations [26-29], self-consistent rate equations [30,31], as well as the nonequilibrium Green's function formalism [32,33]. As was shown above, the quantum cascade lasers (QCL's) are fabricated by stacking up alternating layers of semiconducting material with nanoscale thicknesses. This heterostructure of layers forms a series of conductionband quantum wells in the z direction which trap the electrons into subband states [2] (Figure 1). The calculation procedures described here follows the envelope function approach based on the effective-mass approximation [34,35]. This approximation was found to be much more computationally efficient than atomistic methods, making it more suitable as a design tool for QCLs [36].

The eigenstate of an electron in the unperturbed Hamiltonian of a QCL is the product of the Bloch envelope function $B(x, y, z)$, the free electron wavefunction in the x and y direction, and the bound quantum-well eigen-functions $_n(z)$ in the z direction.

The Bloch function factor contains the effects on the electron state due to the non-uniform nature of the crystal potential on the atomic scale.

We assume the semiconductor layer widths are large compared to the atoms, so we make the approximation that the Bloch function factor is negligible. Each electron is pseudo-free in the x and y dimensions because the material is uniform in those dimensions. Even though each electron is bound to the crystal in these dimensions, we can treat each as free if we use the effective mass of the electron. The bound-state z component wave functions $_n(z)$ are found by numerically solving the one-dimensional Schrödinger equation when the potential profile is known. The potential profile is a combination of the conduction-band edge quantum well profile of the material layers, the bias voltage, and the built-in potential which accounts for the effects of space charge. This minimum energy can be calculated as one of the eigen values of the Schrödinger equation along the growth direction z,

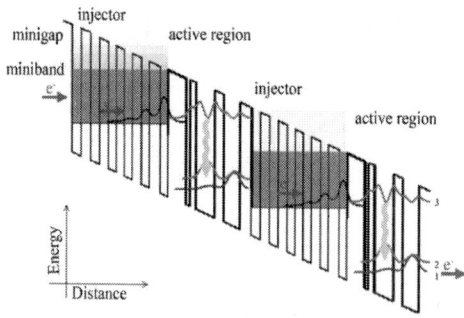

Figure 1: Conduction band profile in two adjacent stages of a generic QCL under an applied bias.

$$\left[-\frac{h^2}{2} \frac{d}{dz} \frac{1}{m_e^*(z)} \frac{d}{dz} + V_c(z) \right] \psi_i(z) = E_i \psi_i(z)$$

(1)

where h is the Planck constant and E_i is the minimum energy of subband i in a QW structure. By solving this equation, one obtains the energy E_i (eigenvalues) and the wave function n (eingenfunctions) of the n electron state. In this method, we have taken into account the boundary conditions for the wavefunction.

An analytical solution provided by solving the Schrö- dinger equation for the conduction minimum energy of subband i has the following form Bas du formulaire

$$E_{cn}(z) = \frac{h^2 \pi^2}{2} \left(\frac{n^2}{m_e^* L_z^2} \right)$$

(2)

Haut du formulaire where L_z represent the depth of the quantum well.

Bas du formulaire.

For a given temperature T, the population n_i for each subband i is expressed by

$$= \frac{m_e}{\pi h^2} k_B T \ln\left(1 + e^{(E_F - E_i)/k_E}\right)$$

(3)

Two simplified expressions can be established for the population n_i:

For the subband situated below the Fermi level $(E_F - E_i \gg k_B T)$, the population is

$$n_i = \frac{m_e}{\pi h^2}(E_F - E_i)$$

For the subband situated above the Fermi level $(E_F - E_i \gg k_B T)$, the population becomes

$$n_i = \frac{m_e}{\pi h^2} k_B T e^{(E_F - E_i)/k_B T}$$

Haut du formulaire The effective states density according to x concentration for the Si/SiGe quantum well is given by

$$\frac{N_{c(SiGe)}}{N_{c(Si)}} \approx \frac{4 + 2 \cdot e^{-\Delta E_{cb}/k_B T}}{6}$$

(4)

where E_{cb} is the limit energy of the conduction band which is differentiated in two terms $E_c(2)$ and $E_c(4)$. $E_c(2)$ is the value linked to both identical directions [001] and [00$\bar{1}$] while $E_c(4)$ is the value linked to the four other identical directions [010], [0$\bar{1}$0], [100] et [$\bar{1}$00].

Bas du formulaire We have then

$$\Delta E_{cb}(x) = E_c(2) - E_c(4)$$

with

$$E_c(2) = 2/3 . \Xi . e_T(x)$$
$$E_c(4) = -1/3 . \Xi . e_T(x)$$

The term $e_T(x)$ is the difference between the strain tensors $e_{zz} - e_{xx}$, according to directions zz and xx. These strain tensors depend of the silicon lattice parameter as well as the SiGe bowing parameter.

RESULTS AND DISCUSSIONS

To design a desired QW structure such as the quantum cascade laser and improve a device performance, a numerical simulation is needed to compute the energy levels and for different electrons states, the corresponding envelope functions, the intersubband transition dipole moments, carrier densities, relaxation times and other parameters. Then, subband formation and energy dispersion are described in the framework of envelope functions with the effective-mass approximation for both conduction and valence band. In Figure 2, we displayed the profiles of the envelope function for different electron states.

In Figure 3, we have illustrated the calculated electronic energy-band structure (a) and total DOS (b) of $Si_{1-x}Ge_x$ alloy for x = 0.5. For the calculations, we have used the full-potential linear muffin-tin orbital (FP-LMTO) method augmented by a plane-wave basis (PLW), implemented in Lmtar code [23-25]. The effects of the approximations to the exchange-correlation energy were treated by the local density approximation (LDA).

Figure 4 shows the variation of the confinement energy with respect to different width wells. This energy is carried out by solving the Schrodinger equation for the Si/SiGe quantum cascade structure. It is clear from this results that for the low width wells the confinement energy is very important. This leads favorably to the intersubband transition. Hence, the emitted wavelength of the QCL only depends on the thicknesses of the layers. One can notice that the use of semiconductor with small effective mass excites well the confinement effects what is not the case for the material SiGe of which its effective mass is important compared to those of GaAs, InAs, equal to 0.067 m_0 and 0.023 m_0, respectively.

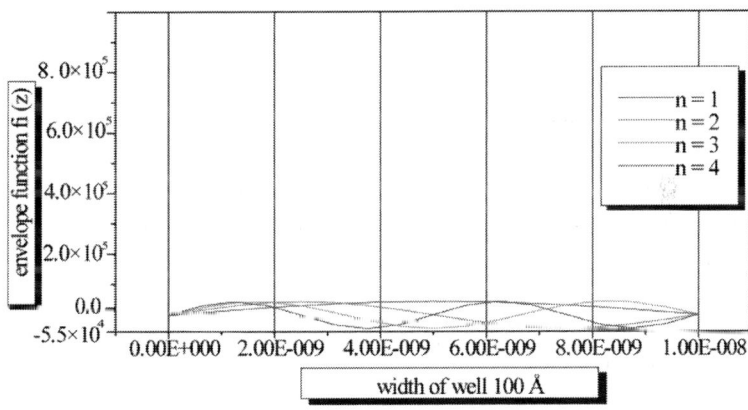

Figure 2: The profiles of the envelope function for different electron states.

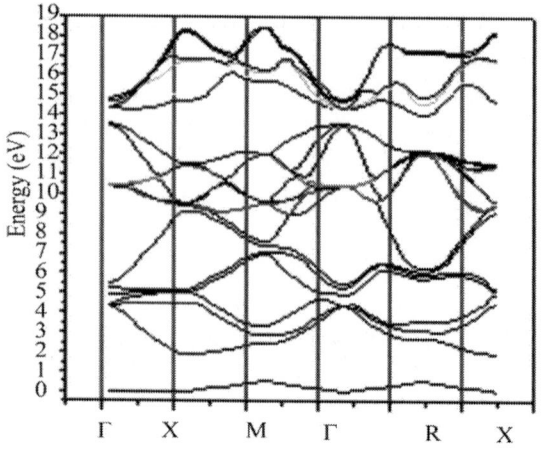

(a)

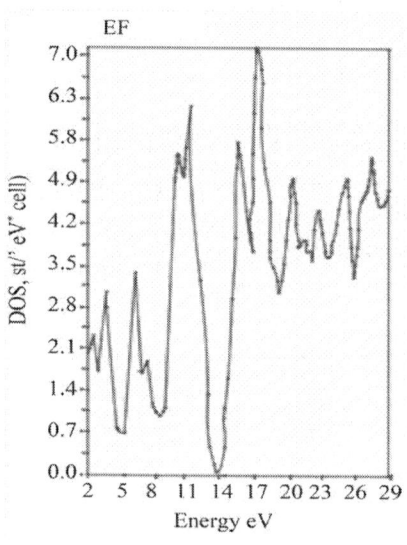

(b)

Figure 3: The calculated electronic energy-band structure (a) and total DOS (b) of $Si_{1-x}Ge_x$ alloy for $x = 0.5$.

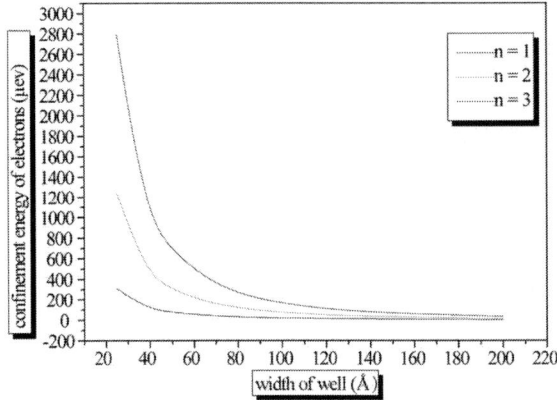

Figure 4: The confinement energy vs the width well for different states.

Figure 5 illustrated the variation of the effective states density with respect to germanium concentration x. We can see that the effective states density of the barrier material is very important to that of the well. Then we shall have continued and a strong carrier's injection in the well to minimize the losses and increase the laser gain.

Figure 6, shows the relative position of minibands for two consecutive transport zones under an applied bias for our quantum cascade structure. By varying the thickness of the wells and barriers of the Si/SiGe super lattice, we modify the position and the size of these minibands. We can extract from this scheme the main functions of the transport zone, which guarantees, the transport of electrons towards the excited subband of the emission zone, the blocking by the miniband gap of the electrons transport of the excited subband towards the continuum, causing then the electron excitations by the inter subband transition. The next step is the injection of the extracted carriers from the fundamental subband towards the next emission zone.

Let us to turn now to the description of the electronic relaxation process with the aim of showing the various rates occurring to describe the intersubband transitions. We consider an emission zone with 3 subbands where the intersubband transition 3 - 2 being radiative. Figure 7 schematizes the various interactions involve for the period where J is the current density, $_3$ is the current injection efficiency in the subband 3 and $_{1,2}$ the proportions of current of flight (leak) in subbands 1 and 2.

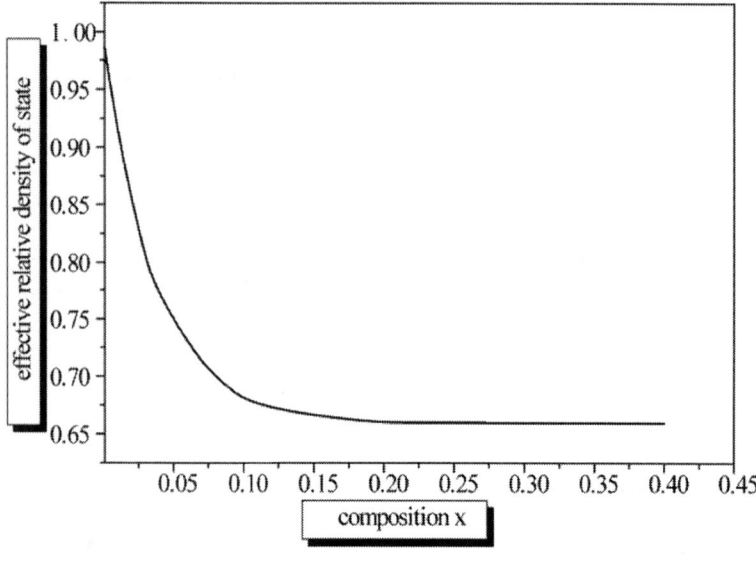

Figure 5: The effective states density vs the germanium concentration x.

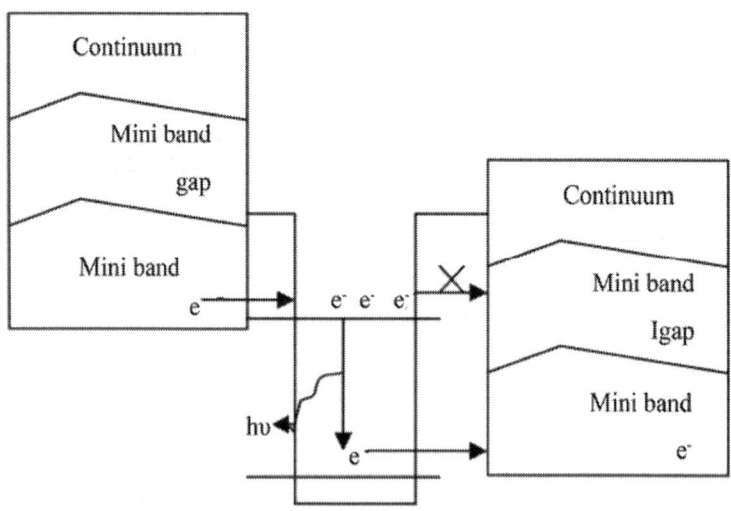

Figure 6: The relative position of minibands for two consecutive transport zones under an applied bias.

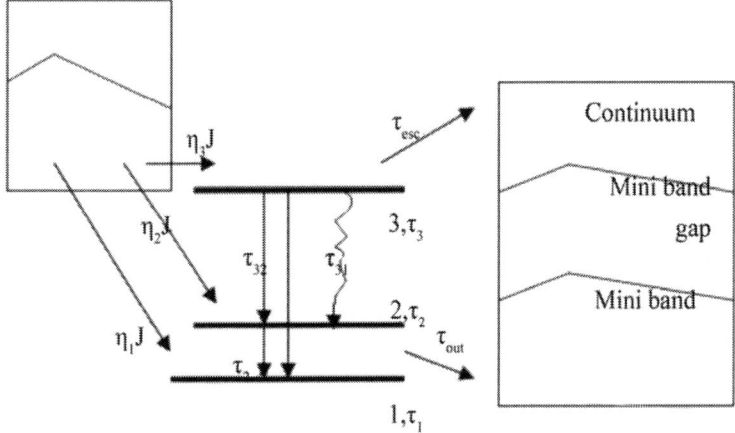

Figure 7: The schematic of the electronic relaxation process.

In the present model, we shall suppose that the injection is completed ($_3$ = 100%). Each subband is associated a life time of electrons noted t_i, i = 1, 2, 3. The Electron-phonon interaction was represented by the rate τ_{ij}^{-1} for an electron transition from the subband i towards the subband j. The electron escape rate towards the continuum was represented by τ_{esc}^{-1}. Finally τ_{out}^{-1} denotes the extraction rate of electrons in the miniband of the following transport zone.

According to these notations [3], the electronic relaxation process is then describes by the following equations:

$$\tau_3^{-1} = \tau_{32}^{-1} + \tau_{31}^{-1} + \tau_{esc}^{-1} + \tau_{spont}^{-1}$$

$$\approx \tau_{32}^{-1} + \tau_{31}^{-1} + \tau_{esc}^{-1} \qquad (5)$$

$$\tau_2^{-1} = \tau_{21}^{-1} + \tau_{out}^{-1}$$

For lasing to occur between two subbands, it is necessary to induce stimulated emission between them. To sustain such emission of photons, there must be sufficient optical gain to compensate various losses in the laser structure. The intersubband optical gain can be obtained by analyzing transition rates between two subbands.

We note that the optimization of the active zone has for an essential purpose to control the electrons phonons interactions, the injection and the extraction of the carriers from the emission zone to the other one and to maximize the strength of oscillator of the radiative transition of this QCL. It is found that the oscillator strength, in the case of our considered system Si/SiGe, corresponding to the intersubband transition between the sub states i and j depends only on the electron effective mass. Hence, a high intersubband optical gain requires a high oscillator strength.

On the other hand, the optical gain is very sensitive to the values of the width well and the doping of the wave guide used in this structure.

In Table 1, we have listed the optimized parameters for our waveguide based on the $Si_{1-x}Ge_x/Si$ system. These results are in good agreement with other calculation [37].

CONCLUSIONS

In this paper, we have carried out the background knowledge necessary to appreciate the quantum cascade laser structure and operation. It begins with a discussion of laser fundamentals and quantum wells and concludes with information on $Si_{1-x}Ge_x/Si$ QC Lasers.

The shape of the confining potential, the sub band energies and their eigen envelope wave functions are calculated by solving a one-dimensional Schrödinger equation. The calculation of optical parameters is used to optimize the Si/SiGe quantum cascade structures. Our results are found to be in good agreement with other calculations. Further works are in progress.

Table 1: The optimized parameters for the $Si_{1-x}Ge_x/Si$ wave guide

Materials	The well width (μm)	Doping (cm⁻³)
Substrat Si n+	1	$\sim 10^{18}$ cm⁻³
Si0.9Ge0.1	1	$\sim 10^{17}$ cm⁻³
Si n	2.5	$\sim 10^{16}$ cm⁻³
$Si_{1-x}Gex/S_i$ (x = 0.3)	1.63	

Si n	2.5	~10^{16} cm^{-3}
Si$_{0.9}$Ge$_{0.1}$	1	~10^{17} cm^{-3}
Si n+	1	~10^{18} cm^{-3}

REFERENCES

1. R. F. Kazarinov and R. A. Suris. "Possibility of the Amplification of Electromagnetic Waves in a Semi-Conductor with a Superlattice," Soviet Physics, Semi-Conductors, Vol. 5, No. 4, 1971, p. 707.

2. J. Faist, F. Capasso, D. Sivco ,C. Sirtori, A. L. Hutchinson and A. Y. Cho, "Quantum Cascade Laser," Science, Vol. 264, No. 5158, 1994, pp. 553-556.doi:10.1126/science.264.5158.553

3. J. Faist, F. Capasso, C. Sirtori, D. Sivco and A. Cho, "Intersubband Transitions in Quantum Wells: Physics and Device Applications II," Academic Press, New York, 2000, pp. 1-83.

4. F. Capasso, C. Gmachl, D. L. Sivco and A. Y. Cho. Physics World, Vol. 12, 1999, p. 27.

5. J. Faist, M. Beck, T. Aellen and E. Gini, "Quantum-Cascade Lasers Based on a Bound-to-Continuum Transition," Applied Physics Letters, Vol. 78, No. 2, 2001, pp. 147-149.

6. H. Page, C. Becker, A. Robertson, G. Glastre, V. Ortiz and C. Sirtori, "300 K Operation of a GaAs-Based Quantum-Cascade Laser at ≈ 9 µm," Applied Physics Letters, Vol. 78, No. 22, 2001, pp. 3529-3531.

7. C. J. Otten, "For Quantum Confinement, Size Matters, But So Does Shape," Washington University, St. Louis, 2003.

8. M. Fukuda, "Optical Semiconductor Devices," John Wiley and Sons, Inc., New York, 1999.

9. F.-Q. Liu, L. Li, L. J. Wang, J. Q. Liu, W. Zhang, Q. D. Zhang, W. F. Liu, Q. Y. Lu and Z. Wang, "Solid Source MBE Growth of Quantum Cascade Lasers," Applied Physics A, Vol. 97, No. 3, 2009, pp. 527-532. doi:10.1007/s00339-009-5423-8

10. C. Sirtori, P. Kruck, S. Barbieri, P. Collot, J. Nagle, M. Beck, J. Faist and U. Oesterle, "GaAs/Al$_x$Ga$_{1-x}$As Quantum Cascade Lasers," Applied Physics Letters, Vol. 73, No. 24, 1998, p. 3486.

doi:10.1063/1.122812

11. F. Capasso, C. Gmachl, A. Tredicucci, A. L. Hutchinson, D. L. Sivco and A. Y. Cho, "High Performance Quantum Cascade Lasers," Optics and Photonics News, Vol. 10, No. 10, 1999, p. 31. doi:10.1364/OPN.10.10.000031

12. C. Gmachl, F. Capasso, R. K. Rohler, A. Tredicucci, A. L. Hutchinson, D. L. Sivco, J. N. Baillargeon and A. Y. Cho, "Mid-Infrared Tunable Quantum Cascade Lasers for GasSensing Applications," IEEE Circuits Devices, Vol. 16, No. 3, 2000, pp. 10-18.doi:10.1109/101.845908

13. C. R. Webster, G. J. Flesch, D. C. Scott, J. E. Swanson, R. D. May, W. S. Woodword, C. Gmachl, F. Capasso, D. L. Sivco, J. N. Baillargeon, A. L. Hutchinson and A. Y. Cho, "Quantum-Cascade Laser Measurements of Stratospheric Methane and Nitrous Oxide," Applied Optics, Vol. 40, No. 3, 2001, p. 321. doi:10.1364/AO.40.000321

14. G. Dehlinger, L. Diehl, U. Genser, H. Sigg, J. Faist, K. Ensslin, D. Grutzmacher and E. Muller. Science, Vol. 290, p. 2277.

15. I. Bormann, K. Brunner, S. Hackenbuchner, G. Zandler, G. Abstreiter, S. Schmult and W. Wegscheider, "Nonradiative Relaxation Times in Diagonal Transition Si/SiGe Quantum Cascade Structures," Applied Physics Letters, Vol. 80, 2003, p. 5371.

16. I. Bormann, K. Brunner, S. Hackenbuchner, G. Abstreiter, S. Schmult and W. Wegscheider, "Nonradiative Relaxation Times in Diagonal Transition Si/SiGe Quantum Cascade Structures," Applied Physics Letters, Vol. 83, No. 26, 2003, p. 5371. doi:10.1063/1.1631381

17. J. C. Phillips, "Bonds and Bands in Semiconductors," Academic Press, New York, 1973.

18. A. R. Jivani, P. N. Gajjar and A. R. Jani, "Total Energy, Equation of States and Bulk Modulus of Si and Ge," Semiconductor Physics, Quantum Electronics and Optoelectronics, Vol. 5, No. 3, 2002, pp. 243-246.

19. T. Soma, "The Electronic Theory of SiGe Solid Solutions," Physica Status Solidi (b), Vol. 95, No. 2, 1979, pp. 427-431. doi:10.1002/pssb.2220950212

20. S. Gonazalez, "Empirical Pseudopotential Method for the Band Structure Calculations of Strained Silicon Germanium Materials," Ph.D. Thesis, Arizona State University, Arizona, 2001.

21. C. J. Williams, "Impact ionization and Auger Recombination in SiGe Heterostructures," Ph.D. Thesis, University of Newcastle, Tyne, 1996.

22. K. Zellat, B. Soudini, N. Sekkal and S. M. Ait Cheikh, "Computational Investigation of Electronic and Optical Properties of Si, Ge, and $Si_{1-x}Ge_x$ Alloys Using the FPLMTO Method Augmented by a Plane-Wave Basis," American Journal of Condensed Matter Physics, Vol. 2, No. 1, 2012, pp. 1-10. doi:10.5923/j.ajcmp.20120201.01

23. S. Y. Savrasov, "Linear-Response Theory and Lattice Dynamamics: A Muffin-Tin-Orbital Approach," Physical Review B, Vol. 54, No. 23, 1996, pp. 16470-16486.doi:10.1103/PhysRevB.54.16470

24. S. Savrasov and D. Savrasov, "Full-Potential LinearMuffin-Tin-Orbital Method for Calculating Total Energies and Forces," Physical Review B, Vol. 46, No. 19, 1992, pp. 12181-12195. doi:10.1103/PhysRevB.46.12181

25. D. Rached, M. Rabah, N. Benkhettou, M. Driz and B. Soudini, "Calculated Band Structures and Optical Properties of Lead Chalcogenides PbX (X = S, Se, Te) under Hydrostatic Pressure," Physica B: Physics of Condensed Matter, Vol. 337, No. 1-4, 2003, pp. 394-403.doi:10.1016/S0921-4526(03)00443-5

26. R. C. Iotti and F. Rossi, "Nature of Charge Transport in Quantum-Cascade Laser," Physical Review Letters, Vol. 87, No. 14, 2001, Article ID: 146603.doi:10.1103/PhysRevLett.87.146603

27. F. Compagnone, A. DiCarlo and P. Lugli, "Monte Carlo Simulation of Electron Dynamics in Superlattice Quantum Cascade Lasers," Applied Physics Letters, Vol. 80, No. 6, 2002, p. 920. doi:10.1063/1.1448664

28. H. Callebaut, S. Kumar, B. S. Williams, Q. Hu and J. L. Reno, "Importance of Electron-Impurity Scattering for Electron Transport in Terahertz Quantum-Cascade Lasers," Applied Physics Letters, Vol. 84, No. 5, 2004, p. 645. doi:10.1063/1.1644337

29. O. Bonno, J. L. Thobel, and F. Dessenne, "Modeling of Electron-Electron Scattering in Monte Carlo Simulation of Quantum

Cascade Lasers," Journal of Applied Physics, Vol. 97, No. 4, 2005, Article ID: 043702. doi:10.1063/1.1840100

30. D. Indjin, P. Harrison, R. W. Kelsall and Z. Ikoni , "SelfConsistent Scattering Theory of Transport and Output Characteristics of Quantum Cascade Lasers," Journal of Applied Physics, Vol. 91, No. 11, 2002, p. 9019. doi:10.1063/1.1474613

31. V. D. Jovanovic, D. Indjin, N. Vukmirovic, Z. Ikonic, P. Harrison and E. H. Linfield, "Mechanisms of Dynamic Range Limitations in GaAs/AlGaAs Quantum-Cascade Lasers: Influence of Injector Doping," Applied Physics Letters, Vol. 86, No. 21, 2005, Article ID: 211117. doi:10.1063/1.1937993

32. S. C. Lee and A. Wacker, "Nonequilibrium Green's function Theory for Transport and Gain Properties of Quantum Cascade Structures," Physical Review B, Vol. 66, No. 24, 2002, Article ID: 245314. doi:10.1103/PhysRevB.66.245314

33. S. C. Lee, F. Banit, M. Woerner and A. Wacker, "Quantum Mechanical Wavepacket Transport in Quantum Cascade Laser Structures," Physical Review B, Vol. 73, No. 24, 2006, Article ID: 245320. doi:10.1103/PhysRevB.73.245320

34. S. R. White and L. J. Sham, "Electronic Properties of Flat-Band Semiconductor Heterostructures," Physical Review Letters, Vol. 47, No. 12, 1981, pp. 879-882.doi:10.1103/PhysRevLett.47.879

35. S. L. Chuang, "Physics of Optoelectronic Devices," Wiley Interscience, New York, 1995.

36. P. Harrison, "Quantum Wells, Wires and Dots," 2nd Edition, Wiley, Chichester, 2005.

37. G. Dehlinger, L. Diehl, U. Gennser, H. Sigg, J. Faist, K. Ensslin, D. Grützmacher and E. Müller, "Intersubband Electroluminescence from SiGe Quantum Cascade Structures," Science, Vol. 290, 2000, p. 2277.

Steady-State Behavior of Semiconductor Laser Diodes Subject to Arbitrary Levels of External Optical Feedback

Qin Zou

Institut Mines-Telecom, Telecom SudParis, Département Electronique et Physique, UMR 5157 CNRS, Evry, France and CEA Saclay Nano-Innov, Gif sur Yvette, France

ABSTRACT

This paper investigates the steady-state behavior of a semiconductor laser subject to arbitrary levels of external optical feedback by means of an iterative travelling-wave (ITW) model. Analytical expressions are developed based on an iterative equation. We show that, as in good agreement with previous work, in the weak-feedback regime of operation except for a phase shift the ITW model will be simplified to the Lang-Kobayashi (LK) model, and that in the case where this phase shift is equal to zero the ITW model is identical to the LK model. The present work is of use in particular for distinguishing the coherence-collapse regime from the strong-feedback regime where low-intensity-noise and narrow-linewidth laser operation would be possible at high feedback levels with re-stabilization of the compound laser system.

INTRODUCTION

Continuous efforts have been made since the last three decades on the dynamics of semiconductor lasers with external optical feedback, because of a large variety of interesting properties they exhibit. A laser with delayed feedback builds an ideal system for analyzing and exploring typical phenomena encountered in a nonlinear time-delayed system, such as bifurcations, thresholds of instability (or stability) and routes to deterministic chaos. On the other hand, external optical feedback can severely affect the spectral behavior of a laser. It can also produce undesirable effects whose effective control is therefore essential for many applications such as optical-fiberbased transmission and sensor systems, since both of them are highly dependent on the spectral quality (temporal coherence and frequency stability) of the used light sources.

In parallel with numerous experimental investigations, theoretical approaches have also been developed aimed at a better understanding of the nonlinear dynamics of a compound laser system (see for example, Tromborg et al. [1], Schunk and Petermann [2], and Binder and Cormack [3]). Most of these approaches have been developed on the basis of the rate equations proposed by Lang and Kobayashi (LK) [4]. For the weak-feedback regime of operation (feedback power ratio less than −30 dB), these approaches have been found to describe adequately various phenomena so far observed experimentally, such as the threshold of coherence collapse (CC), the low-frequency fluctuations (LFFs), and the period-doubling route to chaos [5]. The LK rate equations have also been used to explain the physical mechanisms of the steady-state [6] and transient [7] LFFs, as well as of the chaotic itinerancy for the case of relatively strong feedback [6].

It has been shown that the LK rate equations can be solved analytically by use of asymptotic methods (A detailed description can be found in [8]). In this approach, a laser with weak optical feedback is regarded as a weaklyperturbed nonlinear dynamic system and the threshold of instability corresponds to the first Hopf bifurcation of the LK rate equations. An attempt has been made at interpreting experimental findings of InAs/InP quantum-dash Fabry-Perot lasers by using this approach, such as the onset of CC and the transition from the regime of LFFs to the regime of so-termed fully-developed coherence collapse (FDCC) [9].

Originally, the LK rate equations were proposed to model a single-mode laser with weak feedback and large delays. When the reflectivity of the external reflecting surface is comparable with or greater than the laser facet reflectivity, strong feedback should be taken into account. In this case, the use of the LK model would no longer be justified. Thus, in order to describe the behavior of a feedback laser with arbitrary feedback levels, an iterative traveling-wave (ITW) model was developed [10,11]. By using this model, dynamic and noise properties of a laser subject to strong optical feedback were numerically investigated [12]. The ITW model predicts in particular a significant decrease of the intensity noise in the strongfeedback regime.

More recently, Radziunas et al. used the travelingwave (TW) approach proposed in [13,14] to model a feedback laser, where the system is described by partial differential equations for the electrical fields which counter-propagate along the longitudinal axis of the laser and are coupled through the usual carrier rate equation. A comparison has been made between the LK and TW models, with emphasis on the stability analysis of cavity modes in their continuous-wave states [15].

This paper investigates the steady-state behavior of a feedback laser with arbitrary feedback levels by means of the ITW model. It may be considered as an extension of the works of Langley et al. [12] and Spencer et al. [16]. We provide additional information about the physical insight into a compound laser system and discuss the similarities and the differences between the ITW and LK models. In Section 2, steady-state solutions will be derived for the external cavity modes and compared with previous work. In Sections 3 and 4, a detailed quantitative comparison between the ITW and LK models will be made and the rigorous condition will be given, under which the

ITW model will be simplified to the LK model. Finally, Section 5 will summarize our conclusions.

ITERATIVE TRAVELING-WAVE MODEL

Consider the configuration of Figure 1. A single-longitudinal-mode laser diode is in resonance with an external Fabry-Perot cavity. We assume that r_1, r_2 and r_3 are all real and dispersionless. For this three-mirror system, the dominant resonator is defined by the mirrors with reflection coefficients r_1 and r_3, and multiple round trips inside the external cavity should be in general taken into account for an arbitrary feedback level.

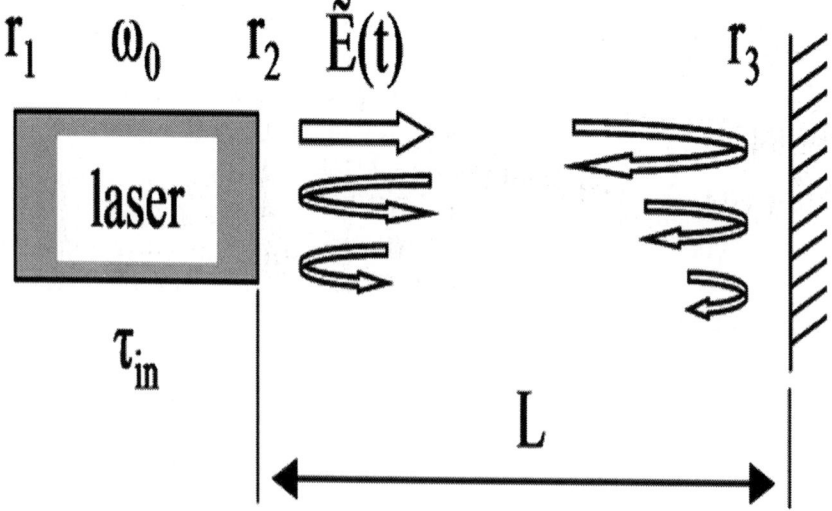

Figure 1: Schematic drawing of a laser diode with external optical feedback. ω_0: emission (angular) frequency of the laser without feedback; τ_{in}: internal round-trip time; r_1: reflection coefficient of the rear facet of the laser; r_2: reflection coefficient of the front facet of the laser; r_3: reflection coefficient of the external mirror; $\tilde{E}(t)$: right-moving electric field passing through the laser front facet; L: length of external cavity assumed empty.

Steady-State Solutions

The right-moving electrical field $\tilde{E}(t)$, calculated at steps of the internal round-trip time $_{in}$ (in seconds) satisfies the following iterative equation [12,16]

$$\tilde{E}(t+\tau_{in})$$

$$= \frac{1}{r_2^2}\exp\left\{\frac{\tau_{in}G_N}{2}(1+j\alpha)\left[N(t)-N_{th}\right]\right\}$$

$$\times\left[\tilde{E}(t)-\left(1-r_2^2\right)\sum_{k=0}^{M}\left(-r_2r_3\right)^k\times\tilde{E}(t-k\tau)\exp\left(-jk\Delta_0\right)\right]$$

(1)

In this equation, G_N (in $s^{-1}\cdot m^3$) is the differential gain; is the linewidth enhancement factor (LEF); $N(t)$(in m^{-3}) is the carrier density and N_{th} its threshold; (in seconds) is the external round-trip time and Δ_0 (in rad, $\Delta_0 = \omega_0\tau$) is the initial feedback phase associated with the emission frequency of the solitary laser operating near threshold.

By inserting $\tilde{E}(t) = \bar{E}(t)\exp(j\omega t)$ into Equation (1) and considering steady-state solutions, we obtain the following expressions for the excess gain ΔG (in s^{-1}) and the feedback phase (in rad, $\Delta= \omega\tau$ with : possible emission frequency) due to the compound structure

$$\Delta G = \frac{-1}{\tau_{in}}\ln\frac{(1-D)^2 + E^2}{r_2^4}$$

(2)

and

$$\text{tg}\left(b_2 - \Delta\frac{\tau_{in}}{\tau}\right)+\frac{E}{1-D}=0$$

(3)

In the above two equations, D (dimensionless), E (dimensionless) and b_2 (in rad) are written respectively as

$$D = \left(1 - r_2^2\right) \sum_{k=0}^{M} \left(-r_2 r_3\right)^k \cos\left(k\Omega\right)$$

(4)

$$E = \left(1 - r_2^2\right) \sum_{k=0}^{M} \left(-r_2 r_3\right)^k \sin\left(k\Omega\right)$$

(5)

and

$$b_2 = \frac{\tau_{in}\alpha}{2} \Delta G = \frac{-\alpha}{2} \ln \frac{\left(1-D\right)^2 + E^2}{r_2^4}$$

(6)

where $\Omega = \Delta_0 + \Delta$

The steady-state behavior of a possible mode (with phase   and excess gain ΔG) produced by a feedback laser under multiple-reflection configuration is described by Equations (2)-(6). We will show in Section 3 that in the case of low feedback levels $r_3 \ll 1$, except for a phase shift, these equations will reduce to the well-employed forms obtained from the LK rate equations. In the following, a cavity mode referred to the LK model will be called an external cavity mode (ECM) and if this mode is referred to the ITW model, it will be called a compound cavity mode (CCM), as suggested in [15].

Initial Feedback Phase

Let us first examine the initial feedback phase Δ_0. With a given system, the phase (or normalized emission frequency)   of a possible mode, being in dependence on Δ_0, is determined through the so-called phase equation. Under the consideration of low feedback levels, the phase equation has a simpler form and two particular situations have been introduced and widely studied, where   is predefined and Δ_0 is then determined from . The first situation corresponds to the "maximum gain mode" defined by the condition $\Delta = 0 (\mathrm{mod}\, 2\pi)$. This gives rise to $\Delta_0 = \gamma\tau\alpha (\mathrm{mod}\, 2\pi)$, where (in s^{-1}) is the feedback rate defined as usual by [1, 2]

$$\gamma = \frac{\left(1 - r_2^2\right) r_3}{r_2 \tau_{in}}$$

(7)

In the second situation, the initial frequency remains unchanged $(\Delta = \Delta_0)$ and the related mode is called the "minimum linewidth mode". We have thus $\Delta_0 = -\mathrm{tg}^{-1}(\alpha)$.

We introduce here another category of modes, where Δ_0 is determined directly from Equation (3). So by putting $r_3 = 0$ and $M = 0$ in Equations (4) and (5), we obtain $D = 1 - r_2^2$, $E = 0$ and, from Equation (6), $b_2 = 0$. Equation (3) becomes then

$$\Delta_0 = \frac{\tau}{\tau_{in}} m\pi \left(m = 0, \pm 1, \pm 2, \cdots\right)$$

(8)

It follows that the change of Δ_0 values is parameterized by the ratio τ / τ_{in}. We note that for both the "maximum gain mode" and the "minimum linewidth mode" there exists the solution $\Delta_0 = 0$ (when $\alpha = 0$), and that this corresponds to the zero-order solution (m = 0) of Equation (8). For simplicity and principle demonstration, we will use in the following $\Delta_0 = 0$ as the value of the initial feedback phase. We will show that in the weak-feedback regime $r_3 \ll 1$ quantitative agreement between the ITW and LK models can be obtained only with this initial phase value.

Excess Gain

A possible CCM has its gain written as

$$G = \tau_p^{-1} + \Delta G$$

(9)

where $_p$ (in seconds) is the photon lifetime and ΔG is the excess gain due to feedback whose expression is given by Equation (2). A contour plot of the evolution of ΔG as functions of and r_3 is shown in Figure 2. As can be seen from this figure, ΔG peaks at the critical point $r_3 = r_2$, which corresponds to a symmetrical (FabryPerot) external cavity.

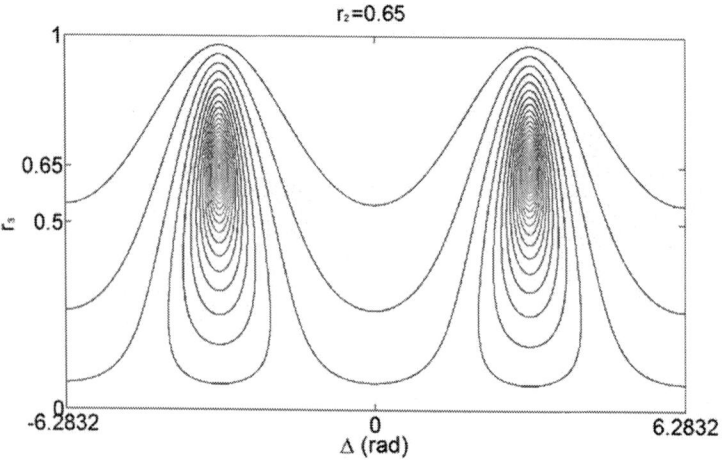

Figure 2: Contour plot of the excess gain G in the vicinity of the initial feedback phase $_0$ ($_0$ = 0), as functions of the CCM feedback phase and the reflection coefficient r_3, for $r_2 = 0.65$, = 5, $_{in}$ = 9 ps, = 0.9 ns, and M = 100.

Modal Density of Photons

From the standard rate equation for the carrier density N [1,4] and by assuming a linear relation between the gain G and N, we can derive the expression for the density of photons (or photon number) I (in m^{-3})

of a CCM. Here I is calculated directly from $\tilde{E}(t)$, we have thus in steady state

$$I = \left|\tilde{E}_0\right|^2 = \frac{I_s - \dfrac{\tau_p}{\tau_s}\dfrac{\Delta G}{G_N}}{1 + \tau_p \Delta G}$$

(10)

where I_s is the density of photons for the solitary laser. It is written as

$$I_s = \tau_p \left(J - \frac{N_{th}}{\tau_s} \right)$$

(11)

where J (in s^{-1}·m^{-3}) is the pumping current and $_s$ (in seconds) is the carrier life time. An example of the typical evolution of I as functions of and r$_3$ is shown in Figure 3.

A typical phenomenon can be observed from this figure: high values of the reflection coefficient r$_3$ do have an effect on enhancement of the density of photons. In this example, I is much higher in the strong-feedback regime than in the moderate-feedback regime (~5 × 10^{20} m^{-3} against ~3 × 10^{20} m^{-3}, which is the threshold value I$_s$ for the solitary laser). As a consequence, the average intensity noise would be decreased in the strong-feedback regime because of higher I values. This result agrees quite well with the work of Langley et al. [12]. They used the ITW model to characterize the transition from the CC regime to the strong-feedback regime for a given r$_2$ value through the relative intensity noise (RIN) (Figure 2(a), in [12]). We see from this figure that with increasing of r$_3$, there are three well distinguished regimes for the RIN: the weak-feedback regime (r$_3$ < 0.005), the noisy CC regime, and the strong-feedback regime (r$_3$ > 0.1). The RIN increases in the CC regime as expected but decreases significantly (more than 10 dB/Hz) in the strong-feedback regime compared to its values in the weak-feedback regime. This result implies that a stable laser operation with low intensity-noise levels would be possible under the condition of strong feedback. In fact, stable and narrow-linewidth operation has been already observed with systems in configuration of strong optical feedback [17]

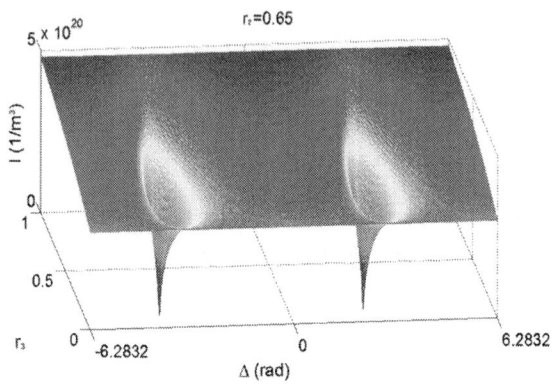

Figure 3: Plot of the density of photons I in the vicinity of the initial feedback phase $_0$ ($_0 = 0$) as functions of the CCM feedback phase ranging from -2ϖ to 2ϖ and the reflection coefficient r_3, for $r_2 = 0.65$. The values of the other parameters are $= 5$, $_{in} = 9$ ps, $= 0.9$ ns, $_s = 2$ ns, $_p = 2$ ps, $G_N = 1.1 \times 10^{-12}$ s^{-1}·m³, $I_s = 3 \times 10^{20}$ m^{-3}, and M = 100.

Another phenomenon is related to the FDCC regime [7]. This regime has been identified for a large pumping current, corresponding therefore to higher output power. In a recent investigation, InAs/InP quantum-dash FabryPerot lasers emitting at 1.57 µm were assessed for their tolerance to external optical feedback by using a freespace setup with a "short" (L = 0.5 m) external cavity [18]. In these experiments, the regime of FDCC was attained for a pumping current of about 30 mA at a rather high feedback level (−1.2 dB in terms of power ratio) with a 600-µm-long laser. As can be seen from the measured RF (radio frequency) spectra reported in this reference ([18], Figure 5(b)), the FDCC regime is characterized by a significant increase of the RF peak power around the relaxation oscillation frequency of the solitary laser, and hence by a high output power in steady state at high levels of feedback. We note that such a behavior manifested by a coherence-collapsed laser can be quite well understood using the formalism developed from the ITW model as can be seen from the RIN spectra simulations ([12], Figure 2(b)), and that the FDCC regime cor responds roughly to the beginning of the strong-feedback regime, as can be seen from the plot of photon density, i.e. Figure 3, in the present text.

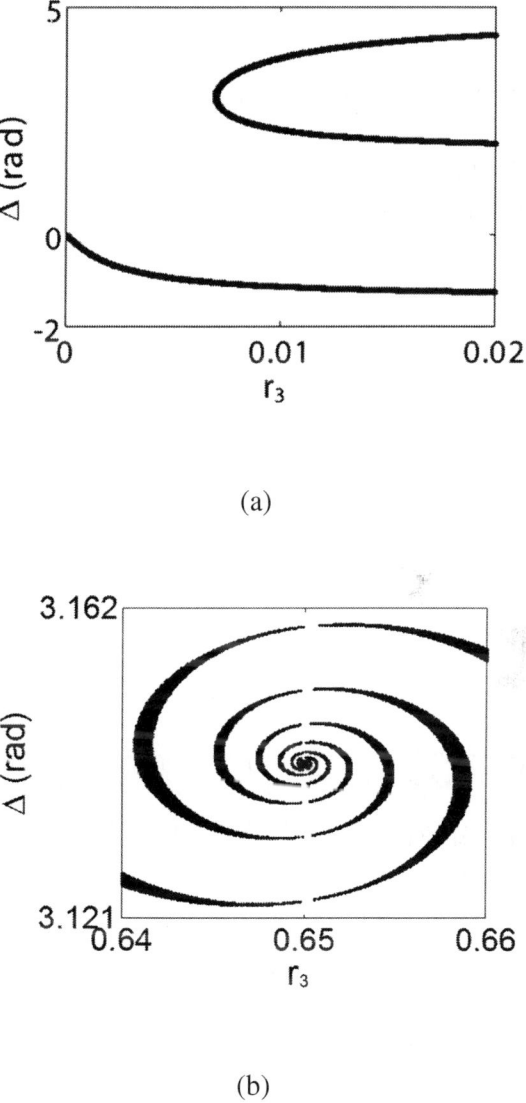

(a)

(b)

Figure 4: Zoom on two particular regions in Figure 5. (a) Region correspond-ing to small values of r_3 (moderatefeedback regime, where the use of the LK model may still be justified); (b) Whirl-shape region near the critical point r_3 = r_2, at which the maximum number of CCMs is found.

Phase Condition

For a given compound-cavity structure, the phase condition determines the emission frequency of a possible CCM. The phase associated with this mode should satisfy the transcendental Equation (3). In general only numerical solutions are possible.

An example of bifurcation diagrams of the CCMs is shown in Figure 5. The value of each point was obtained by a numerical solution of Equation (3). For greater $|\Delta|$ values, it will suffice to repeat the "pattern". We present in Figure 4 two particular regions in Figure 5, where is found in (a) the shape predicted by the LK model as expected. We see from Figure 5 that, when $r_3 < r_2$, same as a classical bifurcation pattern, all the modes (except the first mode Δ_0) emerge by pairs and their number progressively increases with r_3. The maximum number of modes is attained at the point $r_3 = r_2$, corresponding to a symmetrical external cavity. For $r_3 > r_2$, the modes will disappear also by pairs. Finally, the number of modes will become minimal in the strong-feedback regime. We think that Figure 5 is equivalent to the bifurcation diagram for the normalized carrier density illustrated in [12] (Figure 4), showing clearly that high feedback levels can prevent a feedback laser from noisy output.

CONVERGENCE TO THE MODEL OF LANG AND KOBAYASHI

In this section, we will show that in the weak-feedback regime $r_3 \ll 1$, except for a phase term, the ITW model will reduce to the LK model.

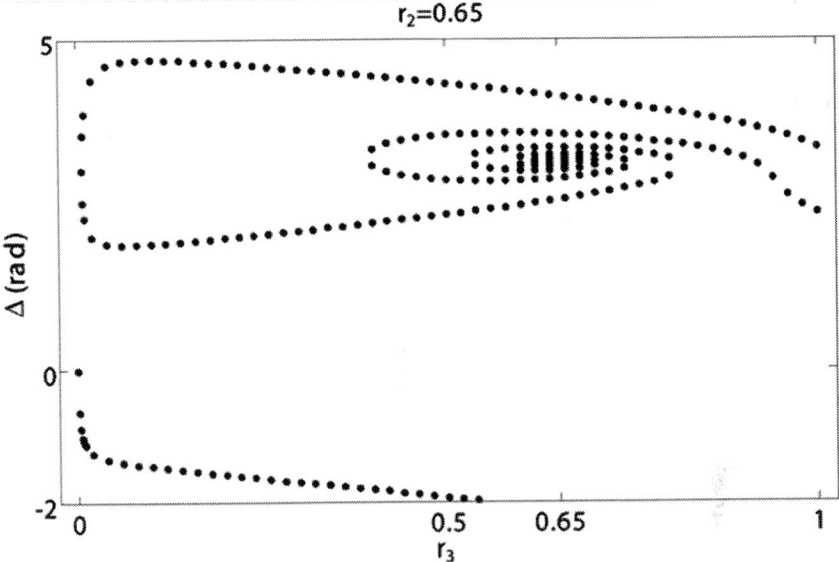

Figure 5: Bifurcation diagram of the CCMs. The values of the parameters are $= 5$, $_{in} = 9$ ps, $= 0.9$ ns, $_0 = 0$, $r_2 = 0.65$, and $M = 100$.

Thus, in the case of low feedback levels, there will be only one reflection ($M = 1$). Equations (4) and (5) are simplified as

$$D = \left(1 - r_2^2\right)\left[1 - r_2 r_3 \cos\left(\Omega\right)\right]$$

(12)

$$E = -\left(1 - r_2^2\right) r_2 r_3 \sin\left(\Omega\right)$$

(13)

Inserting the above two equations into Equation (2) and using the approximations $\left(1 - r_2^2\right) r_3 / r_2 \ll 1$ and $\ln(1 + x) \approx x$ for small x values, we obtain the expression for the steady-state excess gain

$$\Delta G = -2\gamma \cos\left(\Omega\right)$$

(14)

In the same way, for the phase Equation (3), we have

$$\text{tg}\left(b_2 - \Delta\frac{\tau_{in}}{\tau}\right) = \frac{\left(1-r_2^2\right)r_2 r_3 \sin(\Omega)}{r_2^2 + \left(1-r_2^2\right)r_2 r_3 \cos(\Omega)}$$

(15)

Where

$$b_2 = -\gamma\tau_{in}\alpha\cos(\Omega)$$

(16)

By making some approximations and using $\text{tg}(x) \approx x$, we obtain the phase condition

$$\Delta = -\gamma\tau\left[\alpha\cos(\Omega) + \sin(\Omega)\right]$$

(17)

It follows that except for a phase shift Δ_0, Equations (14) and (17) have the same forms as those derived from the LK rate equations. This result confirms the work of Radziunas et al. [15]. They showed a good qualitative agreement between the ECM and CCM solutions at moderate feedback levels. They also found a quantitative agreement for low feedback levels.

DIAGRAM OF THE PHOTON DENSITY VERSUS THE MODE PHASE

It is known that for a laser operating in the weak-even moderate-feedback regime a common way to represent the possible steady states of the ECMs at a fixed feedback level is through an ellipse showing the density of photons I versus the feedback phase , and that only a finite number of ECM points are possible which are all located on the ellipse [8, 19].

For the case of an arbitrary feedback level, the I-Δ diagram with a given pumping current J can be established first by expressing   as a function of ΔG and then by combining the result with Equation (10). We have respectively from Equations (2) and (3)

$$\Delta = \frac{\tau\alpha\Delta G}{2} - \frac{\tau}{\tau_{in}} \text{tg}^{-1}\left(\frac{E}{D-1}\right) \tag{18}$$

and from Equations (14) and (17) for the case of $r_3 \ll 1$.

$$\Delta = \frac{\tau\alpha\Delta G}{2} \pm \frac{\tau}{2}\sqrt{4\gamma^2 - \Delta G^2} \tag{19}$$

Figure 6 shows the plots of the I-Δ diagram for various values of r_3, using respectively Equations (18) and (19). Two typical phenomena inside the CC regime are clearly shown in this figure: a "banana"-like shape and a shift, to positive phase values, of the CCM fixed points due to the inclusion of multiple reflections in the ITW model, as also observed by Spencer et al. when they used the ITW model to establish

the possible steady states in the I versus $\Delta/(2\pi\tau)$ plane ([16],Figure 1). We also observe, in Figure 6, a perfect overlap of the two ellipses for $r_3 = 0.003$. This is because we have taken (for simplicity and principle demonstration) $\Delta_0 = 0$ as the initial phase value.

Finally, let us make a direct comparison between the iteration equation for the ITW model and the rate equation for the LK model. It can easily be shown that at low feedback levels, any stationary solution to Equation (1) will lead to the following equation

$$\frac{G_N}{2}(1+j\alpha)(N_s - N_{th}) + \gamma\exp(-j\Omega) - j\omega = 0 \tag{20}$$

where N_s is the steady-state carrier density. Under the same condition, the LK rate equation will reduce to

$$\frac{G_N}{2}(1+j\alpha)(N_s - N_{th})$$

$$+\gamma\exp(-j\Delta) - j\omega + j\Delta_0/\tau = 0 \tag{21}$$

These two equations show, together with Equations (14) and (17), that the two models are strictly identical if and only if $\Delta_0 = 0$ [see also Figure 6(b) for $r_3 = 0.003$].

CONCLUSIONS

This paper provides additional information about the physical insight into a compound laser system with arbitrary feedback levels. Analytical expressions have been developed based on an iterative travelling-wave model, which enable a characterization in a rigorous way of a cavity mode in its steady state. We show that with decreasing of the reflection coefficient of the external mirror three regimes emerge successively which can clearly be distinguished from bifurcation diagram and gain plot: they are strong-feedback, coherence-collapse, and moderate-feedback regimes. This latter covers the weakfeedback regime where the use of the model of Lang and Kobayashi is entirely justified.

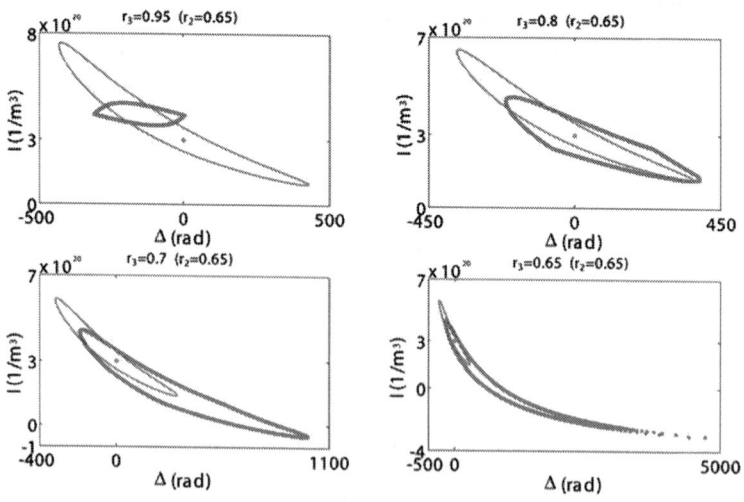

(a)

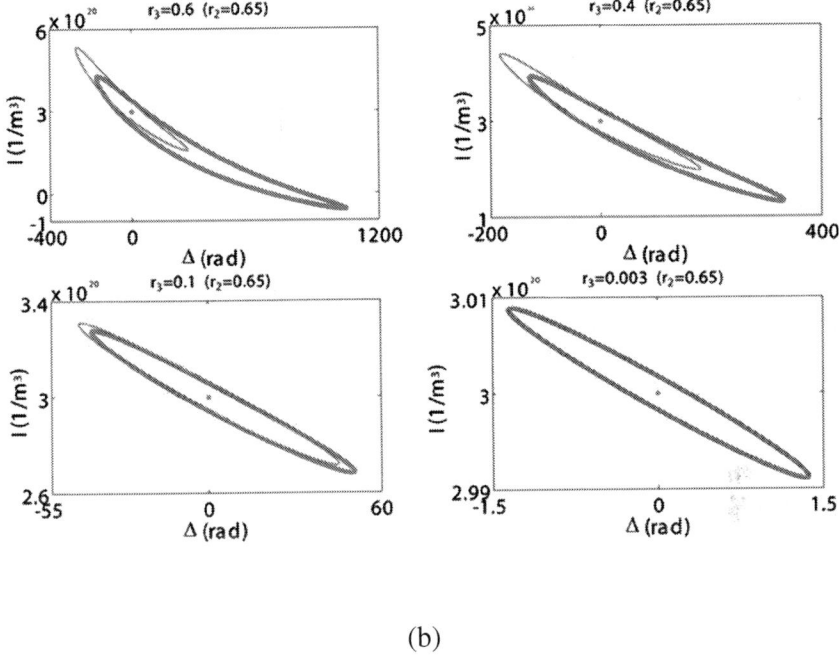

(b)

Figure 6: Evolution of the I − diagram as a function of r_3. The diagrams are determined respectively from Equation (18) (red curves, M = 100) and Equation (19) (green curves). The green dots refer to the solitary-laser solution. (a) r_3 = 0.95, 0.8, 0.7 and 0.65; (b) r_3 = 0.6, 0.4, 0.1 and 0.003. The values of the other parameters are r_2 = 0.65, $_0$ = 0, = 5, $_{in}$= 9 ps, = 0.9 ns, $_s$ = 2 ns, $_p$ = 2 ps, G_N = 1.1 × 10^{-12} $s^{-1} \cdot m^3$, and I_s = 3 × 10^{20} m^{-3}.

We find that maximum number of modes is obtained when the external cavity becomes symmetrical. This state may cause a noisiest laser output. In the strong-feedback regime, a feedback laser is characterized by a minimum mode number and a high density of photons. This behavior confirms previous experimental observations, indicating that beyond the coherence-collapse regime, the system could be re-stabilized and that as a result stable laser operation with low intensity-noise level could be expected with external-mirror reflectivity close to 1. A novel class of modes has been proposed, which is parameterized by the ratio between the external and internal round-trip times. We have also examined the similarities and the differences between the iterative travelling-wave and Lang-Kobayashi models. We find that in the weakfeedback regime these two models are identical only if the initial feedback phase is equal to zero.

The present work is above all useful for determination of the external feedback levels required for stable and noiseless laser output, especially those corresponding to the entrance and the exit of the coherence-collapse regime.

Future investigations will include a detailed analysis of the coherence-collapse regime by means of these two models.

REFERENCES

1. B. Tromborg, J. H. Osmundsen and H. Olesen, "Stability Analysis for a Semiconductor Laser in an External Cavity," IEEE Journal of Quantum Electronics, Vol. 20, No. 9, 1984, pp. 1023-1032. doi:10.1109/JQE.1984.1072508

2. N. Schunk and K. Petermann, "Numerical Analysis of the Feedback Regimes for a Single-Mode Semiconductor Laser with External Feedback," IEEE Journal of Quantum Electronics, Vol. 24, No. 7, 1988, pp. 1242-1247. doi:10.1109/3.960

3. J. O. Binder and G. D. Cormack, "Mode Selection and Stability of a Semiconductor Laser with Weak Optical Feedback," IEEE Journal of Quantum Electronics, Vol. 25, No. 11, 1989, pp. 2255-2259. doi:10.1109/3.42053

4. R. Lang and K. Kobayashi, "External Optical Feedback Effects on Semiconductor Injection Laser Properties," IEEE Journal of Quantum Electronics, Vol. 16, No. 3, 1980, pp. 347-355. doi:10.1109/JQE.1980.1070479

5. J. Ye, H. Li, and J. G. McInerney, "Period-Doubling Route to Chaos in a Semiconductor Laser with Weak Optical Feedback," Physical Review A, Vol. 47, No. 3, 1993, pp. 2249-2252. doi:10.1103/PhysRevA.47.2249

6. T. Sano, "Antimode Dynamics and Chaotic Itinerancy in the Coherence Collapse of Semiconductor Lasers with Optical Feedback," Physical Review A, Vol. 50, No. 3, 1994, pp. 2719-2726. doi:10.1103/PhysRevA.50.2719

7. J. Zamora-Munt, C. Masoller and J. García-Ojalvo, "Transient Low-Frequency Fluctuations in Semiconductor Lasers with Optical Feedback," Physical Review A, Vol. 81, No. 3, 2010, Article ID: 033820. doi:10.1103/PhysRevA.81.033820

8. T. Erneux and P. Glorieux, "Laser Dynamics," Cambridge University Press, New York, 2010.

9. Q. Zou and S. Azouigui, "Analysis of Coherence-Collapse Regime of Semiconductor Lasers under External Optical Feedback by Perturbation Method," Chapter 5, Semiconductor Laser Diode Technology and Applications, Edition InTech, 2012, pp. 71-86.

10. J. Mørk, "Rep. S48," Danish Center for Applied Mathematics and Mechanics, 1989.

11. F. Sporleder, "Travelling Wave Line Model for Laser Diodes with External Optical Feedback," Proceedings of the URSI International Symposium on Electromagnetic Theory, International Union of Radio Science, Brussels, 1983, pp. 585-588.

12. L. N. Langley, K. A. Shore and J. Mørk, "Dynamical and Noise Properties of Laser Diodes Subject to Strong Optical Feedback," Optics Letters, Vol. 19, No. 24, 1994, pp. 2137-2139. doi:10.1364/OL.19.002137

13. J. E. Carroll, J. Whiteaway and R. Plumb, "Distributed Feedback Semiconductor Lasers," Institution of Electrical Engineers, London and SPIE Optical Engineering Press, 1998.

14. U. Bandelow, M. Radziunas, J. Sieber and M. Wolfrum, "Impact of Gain Dispersion on the Spatio-Temporal Dynamics of Multisection Lasers," IEEE Journal of Quantum Electronics, Vol. 37, No. 2, 2001, pp. 183-188. doi:10.1109/3.903067

15. M. Radziunas, H. J. Wünsche, B. Krauskopf and M. Wolfrum, "External Cavity Modes in Lang-Kobayashi and Traveling Wave Models," Proceedings of SPIE, Vol. 6184, 2006. doi:10.1117/12.663546

16. P. S. Spencer, C. R. Mirasso and K. A. Shore, "Effect of Strong Optical Feedback on Vertical-Cavity SurfaceEmitting Lasers," IEEE Photonics Technology Letters, Vol. 10, No. 2, 1998, pp. 191-193. doi:10.1109/68.655354

17. C. E. Weiman and L. Holberg, "Using Diode Lasers for Atomic Physics," Review of Scientific Instruments, Vol. 62, No. 1, 1991.

18. S. Azouigui, B. Kelleher, S. P. Hegarty, G. Huyet, B. Dagens, F. Lelarge, A. Accard, D. Make, O. Le Gouezigou, K. Merghem, A. Martinez, Q. Zou and A. Ramdane, "Coherence Collapse and Low-Frequency Fluctuations in Quantum-Dash Based Lasers Emitting at 1.57 μm," Optics Express, Vol. 15, No. 21, 2007, pp. 14155- 14162.doi:10.1364/OE.15.014155

19. C. H. Henry and R. F. Kazarinov, "Instability of Semiconductor Lasers Due to Optical Feedback from Distant Reflectors," IEEE Journal of Quantum Electronics, Vol. 22, No. 2, 1986, pp. 294- 301. doi:10.1109/JQE.1986.1072959

Citations

CHAPTER 1

Ahmed Bakry, "Modeling of Millimeter-Wave Modulation Characteristics of Semiconductor Lasers under Strong Optical Feedback," The Scientific World Journal, vol. 2014, Article ID 728458, 9 pages, 2014. doi:10.1155/2014/728458.

CHAPTER 2

Moustafa Ahmed, "Theoretical Modeling of Intensity Noise in InGaN Semiconductor Lasers," The Scientific World Journal, vol. 2014, Article ID 475423, 6 pages, 2014 doi:10.1155/2014/475423.

CHAPTER 3

Yu Ye, Lun Dai, Lin Gan, Hu Meng, Yu Dai, Xuefeng Guo, and Guo-gang Qin, Novel Optoelectronic Devices Based on Single Semiconductor Nanowires (Nanobelts), doi:10.1186/1556-276X-7-218.

CHAPTER 4

Faten Adel Ismail Chaqmaqchee, Simone Mazzucato, Murat Oduncuoglu, Naci Balkan, Yun Sun, Mustafa Gunes, Maxime Hugues, and Mark Hopkinson, GaInNAs-Based Hellish-Vertical Cavity Semiconductor Optical Amplifier for 1.3 µm Operation, doi:10.1186/1556-276X-6-104.

CHAPTER 5

Farzaneh Fadakar Masouleh and Narottam Das (2014). Application of Metal-Semiconductor-Metal Photodetector in High-Speed Optical Communication Systems, Advances in Optical Communication, Dr. Narottam Das (Ed.), ISBN: 978-953-51-1730-8, InTech, DOI: 10.5772/58997.

CHAPTER 6

Manuel Durán-Sánchez, R. Iván Álvarez-Tamayo, Evgeny A. Kuzin, Baldemar Ibarra-Escamilla, Andrés González-García and Olivier Pottiez (2013). Experimental Study of Fiber Laser Cavity Losses to Generate a Dual-Wavelength Laser Using a Sagnac Loop Mirror Based on High Birefringence Fiber, Current Developments in Optical Fiber Technology, Dr. Sulaiman Wadi Harun (Ed.), ISBN: 978-953-51-1148-1, InTech, DOI: 10.5772/54330.

CHAPTER 7

Ajay Kumar Yagati, Junhong Min and Jeong-Woo Choi (2014), Electro-chemical Scanning Tunneling Microscopy (ECSTM) – From Theory to Future Applications, Modern Electrochemical Methods in Nano, Surface and Corrosion Science, Dr. M. Aliofkhazraei (Ed.), ISBN: 978-953-51-1586-1, InTech, DOI: 10.5772/57236.

CHAPTER 8

KM Kamruzzaman Selim, Zhi-Cai Xing, Moon-Jeong Choi, Yong-min Chang, Haiqing Guo, and Inn-Kyu Kang, Reduced Cytotoxicity of Insulin-immobilized CdS Quantum Dots Using PEG as a Spacer, doi:10.1186/1556-276X-6-528.

CHAPTER 9

Naveed ul Hassan Alvi, Sajjad Hussain, Jen Jensen, Omer Nur, and Magnus Willander, Influence of helium-ion bombardment on the optical properties of ZnO nanorods/p-GaN light-emitting diodes, doi:10.1186/1556-276X-6-628.

CHAPTER 10

K. Zellat, B. Soudini and S. Cheikh, "Theoretical Investigation of Electronic and Optical Properties of Si/SiGe Quantum Cascade Structures," Advances in Materials Physics and Chemistry, Vol. 3 No. 1, 2013, pp. 19-24. doi:10.4236/ampc.2013.31004.

CHAPTER 11

Q. Zou, "Steady-State Behavior of Semiconductor Laser Diodes Subject to Arbitrary Levels of External Optical Feedback," Optics and Photonics Journal, Vol. 3 No. 1, 2013, pp. 128-134. doi: 10.4236/opj.2013.31021.

Index